IØ126938

Lone Wolf Terrorism Prospects and Potential Strategies to Address the Threat

by
Theodore J. Gordon, Co-Founder of The Millennium Project
Yair Sharan, Director of TAM-C/FIRST and Co-Chair
of the Israeli Node of the Millennium Project
Elizabeth Florescu, Director of Research, The Millennium Project

© 2015 Gordon, Florescu and Sharan
1 Smilax Dr.
Old Lyme, CT 06371

ISBN: 978-0-692-45554-8

2015

Table of Contents

Preface

The world is increasingly aware of Lone Wolf terrorism, but unaware of the future possibilities of a single individual acting alone to make and deploy a weapon of mass destruction. This is referred to as SIMAD (Single Individual Massively Destructive), a term introduced to the public in a Millennium Project scenario *The World Wakes Up*[*].

If one percent of the people in the world in 2050 suffered from psychotic disorders and one percent of them seriously thought about destroying the world, that would be approximately 960,000 people. If one percent of those psychotic people actually tried to make or acquire a weapon of mass destruction in 2050 and succeeded, that would be 9,600 people. If 99% of all those people were caught before they deployed weapons of mass destruction, that would leave 96 successful deployments of weapons of mass destruction by a single individual acting alone. If most of the countries of the world did everything perfectly correct to prevent this future in their country, it could still be irrelevant: the individual(s) could be in countries without a good prevention system. Hence, a global approach is necessary. We have time to think it through and prevent that future.

Prevention could be a three-legged stool:

1. Technical means (e.g., nanotech multi-capacity sensors connected by mesh networks in public spaces, etc.);

2. Collaborations between education and public mental health systems to reduce the number of SIMADs; and

3. Public roles to be created.

During the Cold War, the public had no role in prevention, but can and should have roles in reducing future threats from individuals. This problem is global and multi-faceted, requiring a global and multi-faceted strategy to reduce the threat.

The Millennium Project would like to conduct a study on the public's roles in reducing the likelihood of SIMAD. The technical means of prevention are being created by governments now, and the education/public health roles are better developed by others, but the roles for the publics around the world are not being sufficiently developed—to my knowledge—by anyone. It will take a global organization like The Millennium Project with its 56 Nodes around the world and

---

[*] The Millennium Project, Science and Technology Scenarios, under a grant by the Office of Science, U.S. Department of Energy http://millennium-project.org/millennium/scenarios/st-scenarios.html#Scenario_2._The_World_Wakes_Up_

its collective intelligence system to identify and assess effective global and multi-faceted strategies that include roles for non-classified actors—health services, NGOs, corporations, universities, individuals, media, entertainment industries, religious organizations, and the UN and other international organizations.

Lone Wolf Terrorism Prospects and Potential Strategies to Address the Threat by Theodore J. Gordon, Co-Founder of The Millennium Project; Yair Sharan, Director of TAM-C/FIRST and Co-Chair of the Israeli Node of the Millennium Project; and Elizabeth Florescu, Director of Research for The Millennium Project is an important beginning of a global conversation on how to prevent single individuals acting alone from creating and deploying weapons of mass destruction.

Jerome C. Glenn, CEO
The Millennium Project

Acknowledgments

The Real-Time Delphi study was designed and conducted by Theodore J. Gordon, Co-founder of The Millennium Project, Yair Sharan, Co-Chair of the Israel Node of The Millennium Project and director of the TAM-C FIRST (Foresight Insight Research Science and Technology) group, and Elizabeth Florescu, Director of Research of The Millennium Project. It was produced in cooperation with the Israeli Node of The Millennium Project, FIRST Research Group, and the TAM-C Solutions (former Institute of Terrorism Research and Response).

The final questionnaire was greatly improved thanks to inputs to the pilot project received from Rony Dayan, Michelle Dulaney, Jerome C. Glenn, Jean Negreanu, Freddy Pachis, Aaron Richman, Shelia Ronis, Shlomo Rosenberg, William Tafoya, Charles Thomas, Efrat Tzadik, Gregor Wolbring, John Young, and Simone Di Zio. Jerome C. Glenn, Aaron Richman, and Gregor Wolbring also contributed edits and comments to the final analysis report of the RTD.

The 57 RTD participants (see Appendix C for demographics) provided valuable responses and contributed thoughts and stimulating insights that were followed up in subsequent research.

The authors also acknowledge the valuable comments and legal review provided by Alan J. Kaufman, and Kathleen Housley who reviewed the manuscript and made many valuable suggestions and editorial improvements.

The cover was design by Theodore J. Gordon, using an image he took while flying his glider over Florida.

I think the most likely scenario that we have to guard against right now ends up being more of a lone wolf operation than a large, well-coordinated terrorist attack.
    Barack Obama, President of the United States

Our biggest concern is that a lone individual will unleash a nanotechnology-based doomsday weapon with the same ease as those who send anthrax through the mail. All this individual would have to do is modify technologies that will be easily available to thousands if not millions of individuals by 2020.
    Eric Klein, Founder of Lifeboat Foundation

Synopsis

Trends over the past decades show that Lone Wolf (LW) terrorism is on the rise around the world. To reach a deeper understanding of this phenomenon and its possible future trajectory, the Israeli Node of the Millennium Project conducted a Real-Time Delphi (RTD) study (see Appendix D) in November-December 2013 involving experts in security and related fields from around the world. Results were discussed at a conference in Pescara Italy in June 2014 and at a NATO workshop on Lone Actor Terrorism, in Jerusalem in November 2014[1]. This book is largely based on the outcomes of those works and subsequent research. It presents the results of the RTD study and chapters that focus on key elements of the threats, strategies for coping, and detection of lone wolf terrorism in the future.

The participants in the RTD study thought that a quarter of the terrorist attacks carried out in 2015 might be by a LW. About half of the participants thought that a special category of LW, called single individual massively destructive (SIMAD) might attempt to use weapons of mass destruction by 2030. The number of people killed in a single SIMAD attack was judged to rise over time, although the group's opinion about the possibility of a SIMAD attack killing 100,000 people or more was sharply divided.

The most likely targets were considered to be the "population at large" and "specific population segments". A few respondents pointed to other targets such as agriculture, particularly mentioning a binary anti-crop weapon. The most likely motivations were considered religious incentives and redress of perceived wrongs. "Biotech and synthetic biology" were seen as the most likely fields for the weapons used by potential lone wolves. The most likely location for SIMAD attack(s) was seen as North America (although this might have been influenced by the composition of the panel of participants.)

Three quarters of the respondents agreed that serious attempts to search for a SIMAD would be made before such an attack occurs, and there were several suggestions about what should be done to limit the activities of those identified as potential future lone wolf terrorists. Nevertheless, there was a startling lack of consensus concerning the rate of success of avoiding possible LW attacks.

The study provides some good indications of the areas that deserve more profound research and analysis. The emerging conclusion is that minimizing the LW and SIMAD threats is a long-time continuous effort. Nonetheless, we have an early warning and we should use it!

---

[1] NATO Advanced Research Workshop "Lone Actors—an Emerging Security Threat", organized by TAM2C Solutions and partners; held in Jerusalem, November 3 6, 2014

This book also includes the results of research performed after the RTD study on possible techniques for pre-detection of potential LWs and SIMADs, potential implications of new technologies on the expansion of the scope and spectrum of LW attacks (including the cyber dimension), and some legal and ethical considerations related to addressing the lone wolf phenomenon, as well as some recommended policies for further consideration by security agencies and others likely to be affected or play a role.

We think a new kind of arms race may be developing. On one hand, is the possibility of increasingly destructive weapons falling into the wrong hands, and on the other, the development of new methods of surveillance and pinpointing individuals with malintent. Will the methods of detection be adequate and timely enough to avoid catastrophe?

Throughout the RTD study, the preparation of this book, and the workshop presentations and discussions we consciously avoided giving hints to the bad guys while attempting to impart information that would improve social resiliency and preparedness. In the RTD, the respondents were specifically asked to exclude information that might be considered classified. In addition, the study asked about the general fields from which LW weapons might emerge, rather than specific technologies. The material in this book that is outside of the RTD came from publicly available sources and conferences.

This book includes material derived from several studies on the LW theme done by the co-authors, some having been used in previous publications. The RTD results were published in The Millennium Project's *2013-14 State of the Future* report[2] and have been the basis for an article in *Technology Forecasting and Social Change.*[3] The discussion of new technologies in Chapter 3 was based on results derived of the EU FESTOS project, which focused on new technologies and potential future security threats. Some of these results appeared in an article that was published in *Foresight.*[4] Variations of portions of the material in this book have been presented at conferences and workshops, particularly Chapters 2, 3, and 6 at the "Terrorism and Crime" conference held in Pescara, Italy, in June 2014, and the NATO Advanced Research Workshop "Lone Actors—an Emerging Security Threat" held in Jerusalem, Israel, in November 2014. The material was also used as the basis for a round table discussion at the World Future Society conference in San Francisco, in July 2015.

---

[2] Glenn, J., Gordon, T., Florescu, E., *2013-14 State of the Future*, The Millennium Project, Washington DC., 2014
[3] Gordon, T., Sharan, Y., Florescu, E., "Prospects For Lone Wolf And SIMAD Terrorism," *Technology Forecasting and Social Change*, TFS-18143; February 2015
[4] Hauptman, A., Sharan, Y., "Foresight of Evolving Security Threats Posed by Emerging Technologies," *Foresight*, October 2013

1. Introduction

Lone wolf terrorists have used weapons that range from knives and pistols to machine guns and poisons, from arson to improvised bombs. Their weapons have included aircraft, drones, SUV's and bulldozers. Their motives have ranged from political assassinations to abortion issues, from revenge to religious imperatives, from disagreements with government decisions to simply creating social chaos. Of all forms of terrorism, lone wolf is the most insidious, because it is exceedingly difficult to anticipate, given that it results from the actions and intent of individuals acting alone.

Against this threatening background, the Israeli Node of The Millennium Project[5] (a global futures studies think tank headquartered in Washington DC with Nodes around the world) initiated a study that collected opinions from experts about the outlook for this phenomenon. The participants thought that a quarter of terrorist attacks carried out in 2015 might be by a lone wolf rather than an organized group and that the situation might escalate. About half of the participants thought that lone wolf terrorists might attempt to use weapons of mass destruction by 2030 and although there was sharp disagreement, that 100,000 people might be killed in a single attack by 2033.

In the context of this RTD study, a *lone wolf* terrorist is defined as a single individual acting essentially alone who kills or injures people or inflicts significant damage on essential infrastructure at a single instant or over time, or plans to do so, in order to bring about political, religious, or ideological aims. Targets may be specific groups, undifferentiated masses of people, or infrastructure. This definition is quite restrictive: for example it excludes the Oklahoma City bombing of a Federal Building in April 1995 by Timothy McVeigh because he is believed to have received some assistance from Terry Nichols in carrying out his attack. The accused Boston Marathon Bombers, Dzhokhar and Tamerian Tsamnaev, are excluded as well since two people were involved. It also excludes the horrendous attack on school children and teachers in Newtown Connecticut in December 2012 by Adam Lanza because political, religious, or ideological aims were not involved. (Many other writers would class these events as having been LW acts however, and for this reason they are included in the list of lone wolf episodes that appears in Appendix E.)

Weapons may include firearms, homemade bombs, computers, chemical or biological agents, and weapons of mass destruction (WMD) including self-manufactured as well as military-grade weapons that may have been taken from military sources or government or private laboratories, or purchased on the black market. Cyber terror is also included when it results in massive destruction, or loss of life.

---

[5] The Millennium Project http://www.millennium-project.org

Weapon of Mass Destruction has both a common and a legal definition: commonly, a WMD is any weapon that can kill many people either in an instant (such as a nuclear explosion) or over time (such as a man-made epidemic). Legally, in the U.S., it is "… any weapon that is designed or intended to cause death or serious bodily injury through the release, dissemination, or impact of toxic or poisonous chemicals or their precursors; any weapon involving a biological agent, toxin, or vector; or any weapon that is designed to release radiation or radioactivity at a level dangerous to human life."[6]

There are many known lone wolf incidents, some successful and some thwarted (see Appendix E for some examples.) The weapons that were planned for use or used in these incidents included ricin, anthrax, fire, bulldozers, and other forms of mayhem. Some of the most well-known examples include Anders Breivik, who killed almost 80 people and then surrendered to police in Norway, in 2011; Nidal Hasan, who was tried and found guilty of opening fire at a military base in the United States, in 2009; Ted Kaczynski, the "Unabomber", who despised modern technology and mailed bombs to the scientists whom he apparently thought were developing threatening technologies, between 1978 and 1995; Mohammed Merah, in Toulouse and Montauban, France, who killed seven people including three children in 2012 and was himself killed by police; and Edward Snowden who is accused of stealing and releasing NSA information: his weapon was cyber theft. Recently, in New York City, a LW threat was discovered in time: Jordan Gonzalez, a pharmacy technician pleaded guilty to charges of intent to manufacture ricin, abrin and other dangerous substances at his residences.[7]

A special category of lone wolves is SIMAD: those LWs who use or plan to use WMD. This term was coined in a scenario prepared by The Millennium Project in 2002[8], as part of a study on future science and technology management strategies. It is meant to convey the threat posed by individuals who plan to do harm with weapons that are capable of killing and injuring large numbers of people or destroying or incapacitating infrastructure. In the decade that has passed since the scenario was written, SIMAD became an even more chilling real possibility. One example of a potential SIMAD is the Times Square Bomber, Faisal Shahzad, who was accused of building a bomb from gasoline, propane, and fertilizer and placing the components in an SUV parked in Times Square, and setting the triggering device. Two street vendors discovered the out-of-place SUV and alerted the police. The bomb squad, using remote robots, disarmed it. Shahzad was captured at the Kennedy Airport ready to depart on a flight to Dubai. The indictment listed a

---

[6] 18 U.S. Code § 2332a - Use of weapons of mass destruction. Cornell University Law School, http://www.law.cornell.edu/uscode/text/18/2332a
[7] Zambito T., *Former NJ pharmacist admits developing deadly toxins for a violent confrontation*, NJ.com, May 29, 2014, http://www.nj.com/news/index.ssf/2014/05/nj_pharmacist_admits_developing_deadly_toxins_for_violent_showdown.html
[8] *Future S&T Management Policy Issues: 2025 Global Scenarios*, The Millennium Project, http://www.millennium-project.org/millennium/scenarios/st-scenarios.html#Scenario_2._The_World_Wakes_Up_

number of charges; the first in the list was "Attempted use of a weapon of mass destruction."[9] He was sentenced to life in prison.[10]

A SIMAD's weapon might be a large IED bomb, as it was for Timothy McVeigh but in the future could involve other technologies including biological agents, poisons, or information technology [see Chapters 3 and 4]. Just as weapons might be biological (viruses, alien species), so might be the targets (agriculture, specific population segments.) Driving the growth in this kind of terrorism is an array of new powerful technologies that may become available to individuals. Other technologies are driving techniques for detection of LW plots and terrorists and these include psychological profiling and brain imaging [see Chapter 5]. It is a race to see if the techniques designed to detect and thwart become available and are used in time to avert massive casualties. Some scientists argue that functional MRI (fMRI) and other imaging tools could help to identify potential terrorist inclinations. However, the social implications of searching for potential lone wolves as well as what to do about them if discovered, probably involves compromises of civil liberties and raises controversial ethical dilemmas [see Chapter 6]. So, on the one hand is the possibility of escalation of weaponry; on the other hand, the possibility of new means for early detection of potentially aberrant behavior but possibly at the cost of loss of some liberties.

The Real Time Delphi (RTD) study initiated by the Israeli Node of The Millennium Project was the first step to the wider work presented in this book; it collected judgments from experts to identify the extent of some LW and SIMAD threats and some of the ways to cope with them and, eventually, potential means for countering them.

This RTD study's objectives were to:
- Reach a broader understanding of potential future nature, likelihood and timeframes of LW threats
- Identify technology domains that have the potential of being used by LWs and SIMADs
- Explore policies and approaches concerning the access to information about potentially harmful technologies, as well as to curb their potential impact if they were to be deployed
- Identify possible early warning techniques
- Identify plausible means for identifying potentially threatening individuals
- Assess humane and socially acceptable means for dealing with such individuals
- Collect further ideas and concerns related to these threats

---

[9] Faisal Shahzad Pleads Guilty in Manhattan Federal Court to 10 Federal Crimes Arising from Attempted Car Bombing in Times Square; U.S. Department of Justice, Office of Public Affairs.
http://www.justice.gov/opa/pr/faisal-shahzad-pleads-guilty-manhattan-federal-court-10-federal-crimes-arising-attempted-car
[10] Feyerick, D., *Times Square Bomb Plotter Sentenced to Life in Prison*, CNN.com, October 5, 2010.
http://www.cnn.com/2010/CRIME/10/05/new.york.terror.plot/

The questionnaire used in the study had 17 questions that invited the participants to assess potential timeframes, and likelihood of developments, variables, and strategies associated with LW terror and in addition, potential sources, weapons, and locations, as well as the level of confidence the respondents had in their own answers, and the reasons behind their responses.

The first 13 of these 17 questions were of three types:
1) quantitative- that could be answered with a number such as the expected year of occurrence;
2) multiple-choice questions—to choose from given alternatives; and
3) open-ended—that required write-in answers. The four questions of this type invited the respondents to elaborate on certain issues or suggest other concerns that might be added to the set that appeared in the questionnaire.

The questionnaire also included "mouse over" details that expanded on the definition of a question, historical data where appropriate, Internet links to authoritative articles for additional background, and real-time feedback on the group's responses.

The prospective participants were selected on the basis of their expertise in the areas related to the study. They were personally invited by email. A small pilot group reviewed the first draft of the questionnaire and provided comments and suggestions that were incorporated in the operational questionnaire. A total of 57 people provided responses and valuable inputs; Appendix C shows the demographics of the participants, and Appendix D provides details about the RTD technique.

2. Overview of the Results of the Real-Time Delphi Study

This chapter presents a summary and an analysis of the responses that have come from the RTD; Appendix B presents the full results for each question.

This summary chapter is organized by themes:

2.1 Timeframes and Intensity

2.2 Motivations, Targets, and Location of Violence

2.3 Weapons

2.4 Prevention

2.5 Dealing with Potential Lone Wolves

2.1 Timeframes and Intensity

Following is a table summarizing the responses to the quantitative questions concerning the potential timeframes and intensity (by number of casualties) of an eventual LW or SIMAD attack. Respondents who wanted to answer "never" were instructed to enter the year 2100.

Table 2.1 Potential timeframes and intensity of an eventual LW or SIMAD attack

Question	Min	Max	Average	Median
1. What percentage of terrorist attacks do you think will be carried out by lone wolves in 2015?	0.5	65.0	24.6	25.0
2A. By what year will a lone wolf terrorist kill & injure 500 people?	2000	>2100	2024	2016
2B. By what year will a lone wolf terrorist kill & injure 1,000 people?	2000	>2100	2031	2020
2C. By what year will a lone wolf terrorist kill & injure 5,000 people?	2001	>2100	2047	2027
2D. By what year will a lone wolf terrorist kill & injure 10,000 people?	2017	>2100	2058	2030
2E. By what year will a lone wolf terrorist kill & injure 100,000 people?	2017	>2100	2067	2100
2F. By what year will a lone wolf terrorist kill & injure 1,000,000 people?	2017	>2100	2080	2100
3. When do you think a lone wolf terrorist may first try to use a weapon of mass destruction?	2013	>2100	2033	2020

Some two-thirds of the respondents thought that at least 20% of the terrorist attacks carried out in 2015 will be by lone wolves. The rated average probability was 24.6%, but answers ranged from 0.5% to 65%, with an apparent correlation between the percentage level and the respondent's confidence.

The average year when a lone wolf might use a WMD was judged to be 2033, as noted in the table above, with about half of the participants choosing the interval between 2020 and 2040; confidence levels were generally high.

Using the group's average responses, the number of people killed in SIMAD attacks was judged to rise over time, as shown in Figure 2.1. This may reflect the belief that there will be a growing availability of increasingly destructive technology to LWs. Although there was disagreement, the average group opinion was that a SIMAD attack killing 100,000 people or more could occur before 2050. But half of the respondents felt that such a catastrophe might be much later, if at all. As shown in Figure 2.2, there were no estimates between 2050 and 2075. The respondents' level of confidence in their judgment was mostly "middle".

Figure 2.1 Number of People Killed in SIMAD Attack (average estimated year)

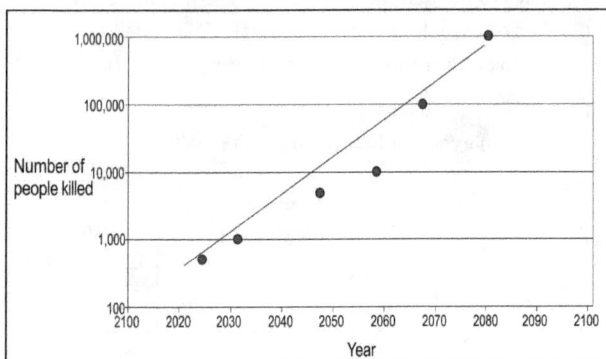

Figure 2.2 Opinions on Timeframes for SIMAD Attack Killing 100,000 or more People

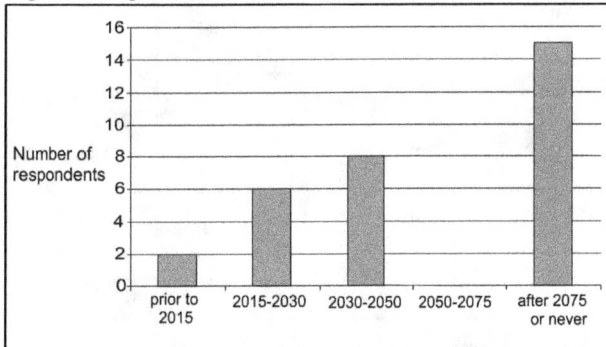

A further analysis of the sharp dichotomy between those who believed that a SIMAD attack might occur before 2050 and those who believed it would occur later on or never, revealed that respondents who identified "security" as their employment saw the threat as occurring much later than those working in other domains. As shown in Figure 2.3, 90% of the participants who self-identified their field as "security" answered that a SIMAD attack killing 100,000 people would occur in or beyond the year 2100 or never.

Figure 2.3 Percentage of Security and Non-security Groups Answering "2100 or Never"

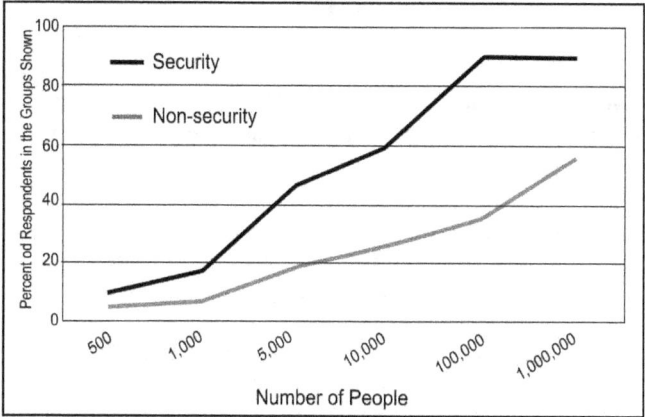

A reason given for the early date of violence was:

> Development of techniques and tools and knowledge... e.g. DIY biotech, synthetic biology; decrease in personal restraint of violent behavior globally, extensive training of violence from Internet, former military, reactions of poor and oppressed used as motivations, lack of spiritual growth in Y-generation and millenials to lead to less self-restraint and hence gravitate to violence.

As for the late date, one respondent said he believed that there would not be a progression from LW terrorism of the current sort to SIMAD killings of massive numbers of people. The respondent said:

> High numbers can be achieved by a LW only with a nuclear bomb which isn't accessible to individuals. A bio epidemic will be stopped quickly.

2.2 Motivations, Targets, and Location of Violence

With a fairly high level of confidence, over 50% of the respondents thought that the most likely location for a SIMAD attack will be North America. Middle East and Europe were second and third rated regions, but with a much lower number of votes, while China and Latin America were not seen to be threatened at all.

The most likely motivations were considered to be religious incentives, and redress of perceived wrongs. The confidence of the respondents was overwhelmingly "high". Psychosis was an additionally suggested motivation.

There was also a high level of agreement that the most likely targets will be "population at large" and "specific population segments". Table 2.2 summarizes the respondents' views concerning the potential location, motivations, and primary targets of LW attacks.

Table 2.2 Potential location, motivations, and primary targets of LW attacks

Questions	Number of Responses						
4. Where do you think an actual lone wolf terrorist attack of the sort described in Question 3 might first occur? [Note: Quest. 3 referred to the use of WMD]	North America	Middle East	Europe	Asia (excl. China)	Africa	Latin America	China
	25	9	8	2	1	0	0
5. However distorted the thinking, what do you think is the single most important motivation that will fuel lone wolf attacks over the next decade?	Religion	Right Wrongs	Insanity	Govern-ment	History	Money	Other
	19	17	3	2	2	1	1
6. What do you think the primary target of lone wolf attacks will be over the next decade?	Population at Large	Population Segments	Infrastructure	Officials	Other	Agriculture/ Food	
	21	16	5	4	1	0	

Commenting about the geographic locations, a respondent noted:
There are strong recent historical reasons for postulating such an attack on a North American (combined U.S.-Canada) target, given the interconnectivity of the grids, but

equally there are strong reasons to believe such attacks could occur elsewhere, particularly in Russia (because of Islamist/khanate region disaffection), and also in the PRC.

Although no one voted for "agriculture" as a potential target per se, one respondent made the following comment:

I have answered population at large, but agriculture runs a close second, particularly if we imagine a binary anti-crop weapon, which can hold harvests at ransom. Terrorists hold the antidote to the poison they have administered to a crop and they extort money or demand actions to allow the harvest.

While another responded noted:

Where government targets are not available, public symbols and social groups will substitute.

2.3 Weapons

The questionnaire asked the participants to indicate the field from which a potential lone wolf would be likely to obtain massively destructive weapons. "Biotech and synthetic biology" received 57.5% of responses, with one respondent repeating that "DIY biology and synthetic biology make development of bioweapons much easier."

"Computers/communications" was a far second, with 12.8% of votes. One respondent commented that "Cyber terrorism is the most likely vehicle for successful, sustained attack by lone wolf."

Confidence was again relatively high, with almost 60% of the respondents having "High" and "Very high" confidence in their responses. Table 2.3 summarizes the responses.

Table 2.3 Number of votes given to fields from which a potential LW would likely obtain WMD

11. Assume that some lone wolf terrorists choose to use massively destructive or disruptive weapons; from what fields might these weapons come?	Biotech/ synthet. biology	Comp./ Commun ication	Agric./ food	Other	Power Generat. and Transp.	Nuclear Physics	Nano tech
	27	6	4	4	2	2	2

There were also some other suggestions including:

Invasive species as a bioweapon; VERY easy to make, distribute, and have severe impact on nations, public health, ecosystems, global food production, and commodity markets.

Nanotech produces agents that are penetrative, saturational, invisible, destructive, and dispersible. Potentially, biggest bang for the buck, even more so than biological agents though nano can be combined with bio.

... very simple weapons that have mass killing implications, such as very effective poisons, distributed in novel ways.

2.4 Prevention

The strategies suggested for prevention could be grouped into two categories: "soft approaches" that include public awareness campaigns and educational reform; and "hard approaches" that include profiling, genetic screening, fMRI brain scanning and other such techniques. Over two thirds of the respondents agreed with a generally high level of confidence that serious attempts to search for lone wolf terrorists who are capable of carrying out an attack using a weapon of mass destruction will be made before such an attack occurs.

Of the technologies likely to be most effective for the detection of people with evil intentions, "Monitoring of purchases of critical materials" and "Monitoring communications and social media" received the overwhelming number of votes, for a total of 71%. (Of course, these techniques are in use today and account for most of the thwarted LW attempts; see Appendix E). Again, the level of confidence ranged overwhelmingly from "middle" to "very high." Table 2.4 summarizes all the responses.

Table 2.4 Technologies likely to be most effective for the detection of people with evil intentions

Question	Number of Participants Choosing a Given Technique						
8. What technology is likely to be most effective for the detection of people with evil intentions? Consider fields such as psychology, brain imaging, observation of unusual behavior, etc.)	Monitor purchases	Monitor commu-nications	Third-party reports of unusual behavior	Mass psycholog. screening	Genetic screening	Brain physiology	Other
	18	14	6	4	3	0	0

The preventive activities seen as most effective were monitoring purchases, monitoring communications, and third party reporting; all of these techniques (along with sting operations) are now being employed and are standard policing techniques. Mass profiling, on the other hand is controversial; one participant suggested such screens might be based on fascination with mass killings, playing violent video games, difficulty in social communications, possibility of being bullied as a child, proficiency with firearms, seizures, lack of emotional connections, and obsessed about battles, war, and destruction. One respondent also suggested searching for a genetic component that would predispose certain people toward this kind of violent behavior, and thus raising the possibility of genetic screening.

However, the responses received to the question about the rate of potential success in avoiding possible lone wolf attacks if the strategies mentioned in Question 8 were implemented, were sharply split. The average response was 53.5%, but ranged from 2% to 100%, as shown in Figure 2.4. The crucial point here is the lack of consensus; but once again, the confidence of the respondents in their answers was fairly high. Table 2.5 summarizes the respondents' views concerning some potential preventive actions.

Figure 2.4. Percent of SIMAD Attacks Averted by using the given set of Strategies

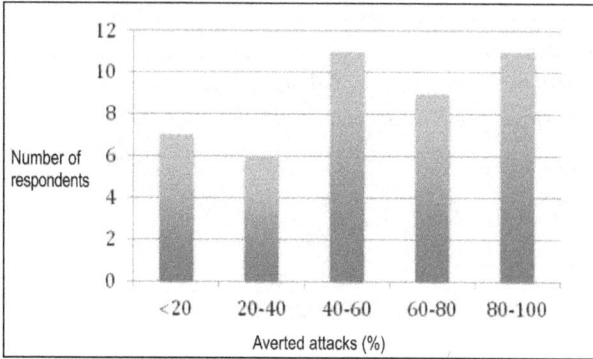

Table 2.5 Views concerning potential preventive actions

Potential Preventive Actions	Responses	
	Yes	No
10. Do you believe that scientific and technological papers and other publications that contain information potentially useful to terrorists should be controlled or withheld?	18	25
12. Are "soft" approaches such as education reform or public awareness campaigns likely to be effective in dealing with the lone wolf threat in the long term?	28	17
13. Are actions which intrude on privacy of people or otherwise compromise their civil rights justifiable in view of the threats?	24	22

The dilemma about the publication of scientific and technological papers that contain information of potential use to the terrorists was reflected again in the split responses of the panel, both groups having a relatively high confidence in their view. However, almost 60% of the participants felt that publication of scientific discoveries should not be withheld. As one respondent said, "because the very nature of scientific research is in its publication."

Similarly, the panel was split over the question of whether or not intrusion into privacy was justified in an effort to limit lone wolf terrorism. On the one hand defending the intrusion based on the potential for preventing mass casualties:

Yes, because of the potential for mass casualties

And on the other, defending the right to privacy:

No. The resolution "Right to privacy in the digital age" adopted by the UN General Assembly in December 2013 is rightly calling on all countries to take measures to end activities such as electronic surveillance, interception of digital communications and collection of personal data, which violate the fundamental "tenet of a democratic society".

While one respondent asked the poignant question:

Who will control the controllers?

About two thirds of the respondents thought that "soft" approaches such as education reform or public awareness campaigns were likely to be effective in dealing with the lone wolf threat in the long term. However, some respondents pointed out some limits:

While soft approaches might work fairly well in most cases, inevitably they will not be 100% effective and therefore the threat will exist still.

And another noted that

such approaches [....] will result, over time, in a smaller pool of people willing to engage in lone wolf terrorism. But unless the global system undergoes dramatic change, there will always be those aggrieved or twisted enough to pursue violence, even if only for its own sake.

Many suggestions were made in response to the open-ended question about steps that should be taken to minimize lone wolf threats. Most supported monitoring, control, education, and intelligence gathering. In addition, there were some new suggestions for mitigating lone wolf threats:

When new-technology weapons are developed, [there is an] obligation to also develop the antidote for them or if not possible, moratorium on the use of the respective technology.

Ban on weaponizable bio/nano technologies. Ban possession of weapons by citizens.

Long term: education and reduction of poverty; higher level public discourse. More immediately: sensors at key transit points and supply sources; focused monitoring of movement of potentially destructive materials - much more effective than data sweeps and tracking every blip in cyber-space.

Minimizing this threat is a long time continuous effort for the national and international authorities that imply economic, social, moral, political, educational factors. Soft and hard measures should be combined according to the specific situations. There is no magic formula to stop this phenomenon.

2.5 Dealing with Potential Lone Wolves

Suppose for a moment that methods were invented to identify people who might become violent lone wolves at some point in the future. How should society deal with them? Some of the respondents warned that the potential fear of large scale violence could lead society down the path to intrusive, ultra authoritarian, and lawless government. Others suggested that due process within the current legal system would be adequate and should be followed. Some two thirds of responses suggest incarceration, removal from society, monitoring, observation or placing on a watch list. The others suggested rehabilitation, counseling, psychotherapy, and medical treatment.

However, one respondent argued:

We cannot lock people up for what they might be thinking absent proof of intent to commit. Monitoring is a balanced approach... If they have not broken any laws, observation. If they have or are about to commit an attack, interception and due legal process.

Another open-ended question asked what if those identified have been engaged in building weapons capable of mass destruction or disruption? How should society treat these potential terrorists? Most of the suggestions fell into the same categories—incarceration, monitoring, and therapy. A new suggestion was to

consider the possibility of special courts/tribunals in which the judges include psychiatrists to examine and prescribe for the suspect.

A respondent commented:

Not enough or not soon enough; will always haunt us. Totalitarian societies seldom have this dilemma.

A final set of open-ended questions asked the participants what steps should be taken to minimize threats of LW and SIMAD respectively, and when. As expected, most of the suggestions overlap. The role of media, education, and social conditions were mentioned the most often both as enhancing factors as well as potential remedial tools:

> *The media should not "heroize" those that commit mass destruction or aggression of any kind.*

> *Media should be full-time involved in sending subliminal messages, so people can reflect about their own behaviors.*

Some of the new interesting ideas were:

> *It is possible that a genetic component exists which induces aberrant behavior. If that is so, genetic manipulation might also be considered. Of course this brings important civil liberty questions to the surface.* [This suggestion is explored in more detail in Chapter 5.]

> *Set up a system where marginally crazy people, under close control and observation, are asked to generate ideas for weapons that a SIMAD might use.*

Some participants made additional comments; a selection includes:

> *Do you think that this threat links to social integration of minorities? What are the early signs of this threat? How can we monitor possible threats? Do you think this threat needs different and new regulations?*

> *Study radicalization phenomenon in general. Societal injustices, perceived threats, deprivations, hurt, actual loss, alienation, sense of loss, sense of pain. Perceptions matter and reality is perceived differently by various people. Group identification of loss of dignity, pride, status, or destiny is important. The emphasis is on group study to get to individual behavior. After all, no lone wolf acts in the blue air. More research effort and dialogue is needed to comprehend and then tackle the issues* [Chapter 6 explores legal and ethical implications].

> *I think more emphasis should be put on lone wolf cyber-terrorists. More questions on cyber crime. We will find the wolves there....* [This is explored in more detail in Chapter 4.]

> *Who is monitoring all the "caretakers" at present and after they retire... who have access to nuclear, medical research, nano etc. materials? The H5N1 controversy is a good example... and I am concerned about "private" science as a whole... who monitors Craig Venter?*

A word should be said about the relatively high confidence that the respondents had in their answers. Many of the events were essentially unknowable, but confidence was generally high. Why? The psychologist Daniel Kahneman says, "high subjective confidence is not to be trusted as an indicator of accuracy (low confidence could be more informative.)"[11] Furthermore, the literature on decision-making and judgment makes the point that experts are often overconfident about their intuitions. [12]

At this point, this RTD study only provides some indications of the areas that deserve more profound research and analysis. The RTD is not a report on the future of LW and SIMAD terrorism, but the aggregate of opinions of a small expert group about what might be. Also, from a methodological standpoint the meaning of the self-evaluation of confidence is unclear.

Nevertheless, the central conclusion emerging from the study is that minimizing the LW and SIMAD threats is a long-time continuous effort for the national and international authorities that imply economic, social, moral, political, educational, psychiatric and other factors. Further research and analysis, as well as building new innovative scenarios would greatly help advance the understanding of the phenomenon and its containment. We have an early warning and we should use it!

---

[11] Kahneman, D., *Thinking Fast and Slow*, Farrar, Strauss and Giroux, New York, 2011
[12]See, for example:
  Angner, E., *Economists as Experts: Overconfidence in theory and practice*,
  http://citeseerx.ist.psu.edu/viewdoc/download?doi=10.1.1.118.9725&rep=rep1&type=pdf
  Kahneman, D., Slovic, P., Tversky, A., *Judgment under uncertainty: Heuristics and biases*, Cambridge University Press. pp. 306–334. ISBN 978-0-521-28414-1

3. The Role of New Technologies: Tools and Targets of Terror

Past terror events show that terrorists are able to control complicated operations using new technologies in order to realize their plans. We envision more and more terrorists and their supporters will have high technical and scientific knowledge. Some of them may be well educated with access to very well-equipped labs that have advanced capabilities. This environment might tempt potential terrorists to act alone and attack people and infrastructure using the advanced technologies available to them. The narrowing gap between civilian and military products and applications make this possibility even more likely. Miniaturization of equipment and systems and cost reduction of military technologies could result in easier availability of weapons to lone actors and facilitate their plans. New breakthroughs in science and technology will become quickly known to terrorists since it is very difficult to control and prevent the dissemination of sensitive knowledge in free societies. Terrorist groups and especially lone actors can use global networks to acquire leading edge knowledge and misuse it for malicious purposes; accelerating technology and its rapid diffusion could place dangerous new options in the hands of potential lone actors and increase security concerns of policy and decision makers as well as society at large.

Yair Sharan, one of the co-authors of this book, was coordinator of a study that was the source of many of the descriptions of future technologies in this chapter and the quantitative charts it includes. The study named "Foresight of Evolving Security Threats Posed by Emerging Technologies (FESTOS)" was a project launched by the European Commission within the framework of their FP security program. FESTOS is one of a series of projects and studies that aims to identify emerging security threats to European society and to find ways to cope with and reduce their impacts.

Specifically, the goal of FESTOS was to identify and assess evolving future threats posed by the abuse or inappropriate use of emerging technologies and new S&T knowledge. A further objective of FESTOS was to evaluate possible ways of limiting and controlling the proliferation of this new knowledge to actors with malicious purposes in order to reduce potential future dangers.

FESTOS was performed in the period 2009-2012 and was conducted using a series of workshops in which the study team and experts participated in a conventional Delphi study; participants were technology and security experts. The FESTOS report appears elsewhere,[13] but key findings from that work are included here.

This chapter describes several fields of science and technology that could result in possible new threats. These include Biotechnology, Nanotechnology, Robotics, and New materials. These fields are rich with promise but have downside applications as well, and of course represent only the tip of the S&T iceberg. Several technology examples are given to demonstrate the dark side of these technologies. We base our descriptions on published sources and data widely available from the FESTOS security study. Being a prospective study, this project looked far into the future, 20 years ahead and more. Some of the technologies considered might be realized in the short or medium range, while others were more futuristic. Security threats resulting from such technologies can serve terrorism, crime and other rouge organizations. Many of the threats identified can serve lone actors as well. Some of the threats can lead to weapons of mass casualties and increase the possibility of such events. The assessments made by experts on the potential likelihood of such threats give a good background to understanding the landscape of future risks and hint about needed counteractions.

3.1 Biotechnology

Biological and biotechnology research covers many activities including classical biotechnology, healthcare (genetics, clinical research and drug development), physiology, synthetic biology, and more. Modern genetic engineering has been instrumental in the rapid development of biological scientific research in the past decades and is expected to be so in the decades to come. Following are a few examples of emerging application areas that might have very positive impacts on health and well being but simultaneously raise future worrisome threats.

Potential applications of all these technologies are foreseen in many areas, including medicine and environment, as well as in the military and security fields and as such they have serious potential for abuse and constitute a threat. The creation or modification of biological organisms poses inherent risks and threats of bio-terror. It could lead to cheaper and widely accessible tools for building bioweapons that could be available to individual terrorists or terrorist organizations.

3.1.1 Synthetic Biology

Synthetic Biology makes use of genetic (and other) materials from biological living forms to design and construct of novel organisms. The main vision is to develop a dynamically expanding

---

[13] Hauptman, A., Sharan, Y., "Foresight of Evolving Security Threats Posed by Emerging Technologies", *Foresight*, volume 15, issue 5 (pp. 375-391) October 2013: http://www.emeraldinsight.com/journals.htm?issn=1463, and Sharan, Y., FESTOS Newsletter, December 2009

---

inventory of standard genomic parts and procedures that engineers can draw from to construct forms with desired functionalities; e.g. manufacturing of vaccines, chemicals and energy forms (e.g. hydrogen, biofuels), coding information, or supplementing human immune systems. Synthetic Biology is inspired from the vision of scientists like Prof. T. Knight from MIT and Prof. Craig Venter who raised the possibility of programming living organisms the same way a computer scientist programs a computer. In this vision, simply saying, interchangeable genetic components--"biobricks"--will be available and their combination will result in a "programmed" organism. Reflecting these visions, European experts defined this emerging technology as: "The engineering of Biology--the synthesis of complex biologically based systems which display functions that do not exist in nature"[14] Practically, this technology opens the way for in vitro building of natural biological agents and new combinations that include artificial agents, from basic building blocks. Bio-engineers will be able, for example, to generate a synthetic genome and then use it to control, or reboot, a recipient cell. Synthetic biology goes beyond classic genetic engineering as it attempts to engineer living systems to perform new functions not found in nature.

Many people, including many scientists and politicians, feel this is an undesirable, immoral, or threatening technology. "Ultimately synthetic biology means cheaper and widely accessible tools to build bioweapons, virulent pathogens and artificial organisms that could pose grave threats to people and the planet," concluded a report by the Ottawa-based ETC Group, one of the advocacy groups that calls for a ban on releasing synthetic organisms.[15] The risk of easy production of new dangerous bio-agents and their possible availability to people with malicious plans will be attractive to potential individuals who would like to initiate an event which will result in a great number of casualties. One can already buy biological laboratories and genetic components such as coding sequences and plasmids online; and informed discussion about whether DIY biology is potentially dangerous has begun.[16]

Terrorist weapons based on these technologies might come from national biological weapons programs. Deadly materials developed in such programs could be stolen by a researcher or lab technician. In the anthrax mailing case the FBI said "charges were about to be brought against anthrax researcher Dr. Bruce Ivins, who took his own life before those charges could be filed."[17] Had Dr. Ivans not first committed suicide, the FBI would have charged him with "the death,

---

[14] Serrano, L., "Synthetic Biology: Promises and Challenges." *Molecular Systems Biology*, Dec 18, 2007. http://msb.embopress.org/content/3/1/158.abstract

[15] Weiss, R., "Synthetic DNA on the Brink of Yielding New Life Forms" for *Washington Post*, December 17 2007. http://www.synbiosafe.eu/uploads/pdf/Washington_Post_Brink_of_Yielding_New_Life_Forms.pdf

[16] Howard Wolinsky, "Kitchen Biology: The Rise Of Do It Yourself Biology Democratizes Science, But Is It Dangerous To Public Health And The Environment?", EMBO Reports, 2009, July: National Institutes of Health: http://www.ncbi.nlm.nih.gov/pmc/articles/PMC2727445/

[17] Anthrax Investigation: Closing a Chapter, The Federal Bureau of Investigation, August 6, 2008. http://www.fbi.gov/news/stories/2008/august/amerithrax080608a

sickness, and fear brought to our country by the 2001 anthrax mailings" that killed 5 people and injured 5 others.[18] Dr. Ivans was a decorated scientist at the US Army's Fort Detrick bio-lab in Frederick, Md. and had access to anthrax spores of the sort used in the mailings. The FBI never had the chance to present their case in court but believed they had strong circumstantial evidence; however, some critics have questioned their findings. Although the case never came to court, it certainly demonstrates the possibility of theft of dangerous material from a site thought to be secure.

3.1.2 DNA-protein interaction

DNA-binding proteins bind to desired sections of a DNA molecule as children's building blocks might fit together; this is a vital function in transcription processes involved in copying genetic information and translating genes into templates for protein production. It is used broadly in applications involving genomic decoding and manipulation and in gene amplification for obtaining large quantities of DNA (billions of copies) from very small samples of DNA or DNA segments. Gene amplification and Polymerase Chain Reaction (PCR) are techniques used in genetic forensics, to diagnose diseases, and to sequence genes.[19] More ominously, these processes can also be used by knowledgable terrorists to develop new infectious agents or virulent forms of known infectious agents, including their weaponization. They might seek, for example, to modify viruses to improvement in their hardiness, level of pathogenicity, and specificity. It is for this reason that the growing hobby of DIY synthetic biology [see Section 3.1.5] has a threatening aspect. Lone wolves with bio-laboratory skills will be dangerous, indeed. But this technology also has a fundamental use in detecting very low levels of pathogens. For example, as mentioned later in this chapter [Section 3.6], Lawrence Livermore National Laboratory has developed a DNA-based sensor capable of detecting over 2,000 viruses and about 900 bacteria using DNA sequencing and PCR.[20]

3.1.3 Biomimicking

Another interesting application is the use of biomimicking to mix fluids at extremely small scale. This application evolves from biomedical research. Scientists plan to speed up biomedical reactions by filling reservoirs with tiny beating rods that mimic cilia. A prototype has been constructed that mixes tiny volumes of fluid or creates a current to move a particle: a flexible

---

[18] Ibid.

[19] Garibyan, L., and Avashia, N., "Polymerase Chain Reaction," Journal of investigative dermatology, 2013 133: http://www.nature.com/jid/journal/v133/n3/full/jid20131a.html

[20] "Researchers at Livermore National Laboratory Develop Microbial Detection Array Capable of Detecting Thousands of Known and Unknown Pathogens in a Single Rapid Test", *Dark Daily*, October 22 2014 http://www.darkdaily.com/researchers-at-livermore-national-laboratory-develop-microbial-detection-array-capable-of-detecting-thousands-of-known-and-unknown-pathogens-in-a-single-rapid-test#axzz3XrxHfGTK, and "Microbial Detection Array", Lawrence Livermore National Laboratory: https://missions.llnl.gov/biosecurity/mda

structure with fingers 400 micrometers long that can move liquids or biological components such as cells at microscopic scale. Such technologies and applications might enable the preparation of toxic substances that need very small scale mixing and are harmful in micro quantities. Toxic materials like these in very small quantities in the hands of a terrorist might pose a difficult threat.[21]

3.1.4 Induced Pluripotent Stem Cells (iPS)

Scientists have used viruses to flip genetic switches in the DNA of skin cells of adult mice to turn them into iPS cells that are functionally equivalent to embryonic stem cells. A recent research showed that mice skin cells can be transformed into neurons quickly and efficiently. According to some opinions, this may enable genetically engineering traits into the cells before using them to create "designer embryos". It also revives concerns regarding reproductive cloning. Dr. Robert Lanza, a stem cell researcher said: "With just a little piece of your skin, or some blood from the hospital, anyone could have your child--even an ex-girlfriend or neighbor.... This isn't rocket science--with a little practice, any IVF clinic in the world could probably figure out how to get it to work."[22] There is a potential that such techniques will be used by criminals or "rogue organizations" for malicious purposes. Which leads to the question about whether there could be designer terrorists?

3.1.5 Gene Transfer

New devices and methods are being developed for transferring genes from one living organism to another. What genes would terrorists want to obtain or want to interfere with? Of course this is pure speculation and if at all plausible, far in the future, but possibly: the preservation and transfer of some genetic properties of spiritual or charismatic leaders to progeny, great physical prowess, improved mental agility, etc., and the opposite for their adversaries. The basic technology will be increasingly available (and affordable) in the future. "Bio-hacking" could happen in the near term: it would be to biology what computer hacking is to computers and automation. A brave new world for criminals and terrorists, and builders of future cultures.

Imagine some other possibilities in this field. It will be possible to type out genetic codes like computer codes and insert the sequences into living organisms in a relatively simple laboratory. Karen Wientraub writing in the journal *Technology* wonders, "Will we all be tweaking our

---

[21] Hickey, H., "Stirred, Not Shaken: Bio Inspired Cilia Mix Medical Reagents At Small Scale", University of Washington, June 30, 2009. http://www.washington.edu/news/2009/06/30/stirred-not-shaken-bio-inspired-cilia-mix-medical-reagents-at-small-scales/

[22] Stein, R., "Researchers Create Cells That They May Be Equivalent To Embryonic Stem Cells", *Washington Post*, July 24, 2009. http://www.washingtonpost.com/wp-dyn/content/article/2009/07/23/AR2009072301786.html

genetic codes?"[23] And in an age of computers that are intelligent that question might become "Could some autonomous computer build the genetic code for an ultimate terrorist or even an ultimate terrorist detector?"

In the shorter term, synthetic biology is spreading as a hobby--from science fairs to home labs. Even today, hobbyists can buy biological components online, and some people are modifying genetic properties and creating chimeras to form hybrid plants and insects; e.g. plants that glow-in-the-dark. Some experts say DNA codes can be written like computer codes. It seems reasonable to expect that some of these hobbyists will be inclined to build helpful material like antibiotics that can overpower antibiotic-resistant super bacteria, but some will be potential LWs and will try to build organic materials such as weaponized bio toxins and viruses that target specific groups.[24] And if they have a little talent, even today there are gene synthesis services online--one can order gene sequences from their keyboard.[25]

3.1.6 Personalized Medicine

In personalized medicine, therapies are customized to patients; for example, genetic biomarkers can predict how a particular patient will respond to a particular drug. Important applications already exist in the treatment of blood clots, colorectal cancer, breast and ovarian cancer, melanoma, and cardio vascular disease; certainly other diseases that come from mutations in DNA--perhaps even Alzheimer's will be included. It is now possible to obtain a personal biological assay--that is an individual's DNA sequence from several private companies at relatively low cost.[26] While drugs are assumed to work well in the general population and they usually do, individual differences make drugs more or less effective in some individuals. In personalized medicine, those differences are paramount. Pharmacogenomics is the study of relationships between human genetics and drug effectiveness. This body of knowledge will make the development of new drugs (personalized drugs) possible, based on individual differences between people and their responses to drugs (such as their ability to metabolize drug products). DNA analysis can also lead to identification of propensity to certain diseases such as various forms of cancer. The actress Angelina Jolie said she carries the BRCA1 gene mutation, a genetic indicator of increased risk of breast and ovarian cancer, and as a result she elected to have prophylactic mastectomies and removal of her ovaries and fallopian tubes.[27]

---

[23] Weintraub, K., "Will We All Be Tweaking our Genetic Codes?" *Technology*, Sept 18, 2011. http://www.bbc.com/news/technology-14919539

[24] Wadhwa, V., "DNA: The Next Big Hacking Frontier", *The Washington Post*, Dec 8, 2011. http://www.washingtonpost.com/national/on-innovations/dna-the-next-big-hacking-frontier/2011/12/07/gIQAmd2KdO_story.html

[25] Genewiz: http://www.genewiz.com/public/gene-synthesis.aspx

[26] For example, 23andMe offers DNA analysis kits for $99. https://www.23andme.com/en-ca/

[27] Angelina Jolie: "I got my ovaries removed: I'm on a mission to live" TMZ, March 24, 2015. http://www.tmz.com/2015/03/24/angelina-jolie-ovaries-removed-cancer/

Gene therapy is a general approach to the treatment of genetic disorders by delivering functional 'working genes' into the body to replace missing or malfunctioning genes. The working genes may be either natural or engineered genetically through recombinant DNA; their usual intent is to induce the body to produce a missing protein and thus avoid or cure a disease.

When genetic material in a sex cell is modified, the intent is to achieve a particular trait or remove an inherited disease or abnormality in subsequent generations. This application is often called "designer babies." BGI, a Chinese company specializing in DNA sequencing has a program underway to identify the genetic basis for intelligence; it is collecting and analyzing DNA from 2,000 people with IQs over 150.[28] If this is successfully accomplished, there is a chance that it could lead to the ability to select for this trait through screening of embryos or zygotes and subsequently implanting only selected organisms. This is a form of eugenics, and is generally considered taboo and by some, bad science because it is likely that many genes are involved in intelligence and the role of environment is ignored by this approach. However, Geoffrey Miller, an evolutionary psychologist at NYU is quoted a saying: "Even if it only boosts the average kid by five IQ points, that's a huge difference in terms of economic productivity, the competitiveness of the country, how many patents they get, how their businesses are run, and how innovative their economy is."[29]

What does all this mean for LW terrorism? There are several implications. The genetic screening being performed to identify genetic analogues to high IQ could also apply to screening for tendency toward violence [see Chapter 5 for a discussion of the potential role of the MAOA gene]. By identifying genetic differences among races, a field also considered taboo, conceivably genetic warfare could be waged against specific races sometime in the distant future. Genocide would take on new meaning. To the extent that DNA information about individuals is stored in databases, unauthorized intrusion could compromise privacy and manipulation of the data could confuse diagnoses and result in inappropriate therapies.

Finally, all of the techniques described here also apply to agriculture. Modern biotechnological methods have been instrumental in the development of new genetically-modified crop varieties with improved yields, reduced vulnerability to environmental stresses and resistance to pesticides and herbicides. Biotechnology has also been applied in the production of novel non-food substances such as detergents, biofuels, drugs and vaccines production by genetically modified plants and more. There have been past terrorist attacks on food and food supply around the

---

[28] "Designer Babies On The Way? In China, Scientists Attempt To Unravel Human Intelligence," *CBS News*, March 5, 2014. http://www.cbsnews.com/news/designer-babies-on-the-way-in-china-scientists-attempt-to-unravel-human-intelligence/

[29] Eror, A., "China is engineering genius babies" *Vice.com*, March 15 2013: http://www.vice.com/read/chinas-taking-over-the-world-with-a-massive-genetic-engineering-program

world. The Global Terrorism Database (GTB) shows a growing number of terrorist attacks on food or water supplies since 1970, reaching over 50 in 2013, the latest year at the time this database was examined.[30] Although most of the events reported were the result of bombings near food or water facilities, several were poisonings. In one of these, reports the GTB, members of the Rajneeshee group attempted to influence an election by introducing salmonella into restaurant salad bars to poison voters and keep them away from the polls.[31] This happened in The Dalles, Oregon, in 1984 and over 700 people were sickened.

Researchers at the RAND Corporation have studied the threat of bioterrorism against livestock.[32] After making the case that agricultural and food industries are vulnerable to attacks and disruptions because they are highly concentrated, have insufficient security, a declining pool of veterinarians and other specialists, and because owners of agricultural facilities are reluctant to report breeches in security that may affect their ability to continue operations, they say:

> "...the capability requirements for deliberately exploiting these weaknesses are not significant. Not only are there a large number of potential pathogens to choose from, but many of these microbial organisms are highly transmissible (something that is particularly true of foot and mouth disease, or FMD) meaning that there is no obstacle of weaponization to overcome. Moreover, because most livestock diseases cannot be passed to humans, there is no requirement on the part of the perpetrator to have an advanced understanding of animal epidemiology, nor is there any need for elaborate containment procedures and/or personal protective equipment (PPE) in the preparation of the agent."[33]

The likelihood and threats of some of these technologies were assessed by technology experts who participated in the FESTOS study. The results are presented below in Fig 3.1. In this and other charts that follow in this chapter, likelihoods of technologies to pose security threats were rated on a scale of 1 to 5 as function of time from present until after 2035. In their judgment, most of the bio-threats discussed in this chapter will mature only after 2020. The likelihood of the threats increases with time, hinting at a continuing progression of the ability of terrorists to adapt the technologies for malicious purposes. If perpetrators overcome the difficulties and realize the threats, mass casualties as well as bizarre attempts at genetic manipulation might be part of the scenario. As Figure 3.1 shows, the perceived likelihood of threats continues to

---

[30] Global Terrorism Database, The National Consortium for the Study of Terrorism and Responses to Terrorism (START), University of Maryland:
http://www.start.umd.edu/gtd/search/Results.aspx?expanded=no&search=food&ob=GTDID&od=desc&page=1&count=50#results-table
[31] GTB and "Targets for Terrorism: Food and Agriculture" Council on Foreign Relations, January 1 2006.
http://www.cfr.org/homeland-security/targets-terrorism-food-agriculture/p10197
[32] Kelly, T., et al., *The Office Of Science And Technology Policy Blue Ribbon Panel Of The Threat Of Biological Therapist Directed Against Livestock*, RAND Corporation, April 2004.
http://www.rand.org/content/dam/rand/pubs/conf_proceedings/2005/CF193.pdf
[33] Ibid.

---

increase into the far future. Threats emerging from Synthetic Biology technology as well as Gene Transfer were seen as more likely than others. It should be mentioned that misuse of genetic technologies might also affect the ethical and value systems: what seems an acceptable risk today, might well be unacceptable in the future.

Figure 3.1 Likelihood of Bio-technologies to pose a security threat

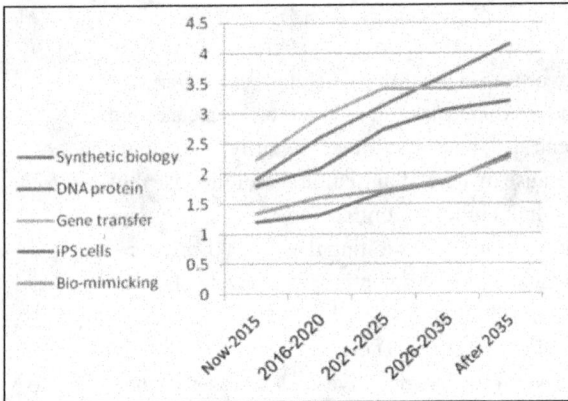

3.2 Nanotechnologies

Nanotechnologies deal with phenomena at nanoscale and the ability to manipulate matter at that scale, namely, at atomic, molecular and macromolecular levels (1 nanometer = 10^{-9} meter). At the nanoscale the borders between classical disciplines (e.g. physics, chemistry, biology) blur, and material properties (e.g. optical, magnetic, electronic, chemical, mechanical) are strongly influenced by any change in size of the material elements under consideration.

Advances in nanotechnology (or NT) have led, and are likely to continue to lead, to many beneficial applications in diverse areas. These include, for example, super-strong lightweight materials, highly efficient photovoltaic and fuel cells, the further miniaturization of electronic components, ultra-dense magnetic discs, targeted drug delivery systems, direct connection between electronics and biology (e.g. electronic devices connected to nerve cells), biocompatible implants, nano-electromechanical systems (NEMS), and many more. More futuristic NT visions include molecular assemblers and self-replicating nanobots. First generations of nanotech-based products are already in use and their markets are expected to grow very significantly in the coming decade. The European Commission expects that they will significantly contribute to the European Union's growth and competitiveness and to the realization of many of its policies, including industry, innovation, environment, energy, transport, security, and space.

The Nobel Laureate Richard Smalley, testifying before the U.S. Congress in a hearing on nanotechnology on June 22,1999, declared that "The impact of nanotechnology on health, wealth, and lives of people will be at least the equivalent of the combined influences of microelectronics, medical imaging, computer-aided engineering, and man-made polymers developed in this [the 20th] century." [34]

Nevertheless, NT is accompanied by various safety risks and hazards and also of intentional abuse. Numerous studies have identified NT-related risks, many of them associated with current industrial research and in particular with defense R&D. Naturally, if these technologies become more easily accessible and affordable, the risk potential may grow.

Examples of existing and potential threats include health and environmental risks of nanoparticles, dangerous result of nano-enabled genetic manipulations, explosives and propellants enhanced by nanoscale additives, NT-enabled active camouflage, networks of ultra-miniature tracking sensors ("smart dust"), and even nanotechnology-based-doomsday weapons based on futuristic nano-assemblers or self-replicating nano-robots.[35]

3.2.1 Nanoassemblers

"Universal nanoassemblers" are an important potential development in molecular nanotechnology. According to this vision, with the appropriate program, the assemblers could autonomously assemble and manufacture almost any desired product "bottom up", molecule by molecule. But they could also self-replicate and this process could continue to produce exponentially increasing numbers of machines. Self-replicating nano-assemblers or nano-robots as an outcome of advanced molecular nanotechnology, and related "dark side" scenarios, have been described in the Science Fiction literature and in several speculative articles, and have been often dismissed by scientists as unrealistic. The idea was introduced for the first time in 1959 by Richard Feynman[36] and the nanotech visionary Eric Drexler (sometimes called "the father of nanotechnology") discussed it in his influential book "Engines of Creation" (1986). He also speculated about the so-called "Grey Goo Scenario": uncontrolled "runaway self-assembling nanoreplicators" that cause damage, kill and destroy the world. Intentional design of such devices once this technology is realized could thus lead malicious use if safeguards fail. But recently, Phoenix and Drexler have downplayed this threat:

---

[34] House Hearing, 106 Congress, *Nanotechnology: The State Of Nano Science And Its Prospects For The Next Decade*; Hearing before The Subcommittee on Basic Research of the US Congress, U.S. Government Printing Office, June 22, 1999. http://www.gpo.gov/fdsys/pkg/CHRG-106hhrg60678/html/CHRG-106hhrg60678.htm
[35] See for example: *Dangers of Molecular Manufacturing*, Center for Responsible Nanotechnology. http://www.crnano.org/dangers.htm
[36] Feynman, R., *There's Plenty Of Room At The Bottom*. http://www.zyvex.com/nanotech/feynman.html

---

"Runaway replicators, while theoretically possible according to the laws of physics, cannot be built with today's nanotechnology toolset. Self-replicating machines aren't necessary for molecular nanotechnology, and aren't part of current development plans.... Runaway replication would only be the product of a deliberate and difficult engineering process, not an accident. Far more serious, however, is the possibility that a large-scale and convenient manufacturing capacity could be used to make powerful non-replicating weapons in unprecedented quantity, leading to an arms race or war. Policy investigation into the effects of molecular nanotechnology should consider deliberate abuse as a primary concern, and runaway replication as a more distant issue."[37]

3.2.2 Nano-Weapons

The threat of non-replicating nano weapons mentioned by Phoenix and Drexler in the previous quotation is of concern: new kinds of weapons or hazardous materials may be produced in small "nanofactories". Some possibilities that have been discussed in the open literature include:

- Energetic nanomaterials are additives that can improve the chemical reactivity and other properties of materials. For example, nano-sized aluminum can be highly explosive. Entirely new molecules with high energy density could be created by molecular nanotechnology methods resulting in new powerful propellants and explosives.[38]

- Nanoparticles can enter the body by ingestion, inhalation or via the skin and could be hazardous to health. Experiments at the University of California found that small iron oxide particles stunted the growth of nerve cells and that short nanotubes could "interfere with human lung cells."[39] Some particles can cross the blood/brain barrier and this property can be exploited in developing new drug delivery systems. Yet, on the whole, the ease by which nanoparticles can enter the body and do harm suggests they may be attractive to a terrorist who wants to cause harm to a great number of people, a SIMAD weapon.

- A third example of a nano weapon concept is of insect spy drones that operate singly or in swarms and are/is remotely controlled or autonomous and can collect information or samples, plant trackers, or sabotage.

---

[37] Phoenix, C., and Dressler, E., "Safe exponential manufacturing", *Nanotechnology*, volume 15, number 8: http://iopscience.iop.org/0957-4484/15/8/001
[38] Lebret, B., Energetic nanomaterials: towards technological breakthrough, CLEFS CEA No 52, Summer 2005. http://www.docstoc.com/docs/93119556/Energetic-nanomaterials-towards-technological-breakthrough
[39] Johnson, R. C., "Studies Warn Of Nano Particle Health Effects", *EE Times*, April 13 2007. http://www.eetimes.com/document.asp?doc_id=1165860

3.2.3 Medical Nanorobots

Robots are now commonly used in surgery; they are operated by surgeons outside of the body in orthoscopic surgery and elsewhere and are micro (not nano) scale machines. Nanorobots, however, are envisioned as nanoscale machines to be inserted into and function in the human body, for example to identify toxic chemicals and their location or to identify cancer cells and deliver drugs to their site. A possible "shortcut" to bio-nanorobotics is to engineer natural nanosystems (e.g viruses and bacteria) to create new, artificial bio-devices. Future advances could include controlled swarms of molecular nanorobots, reacting faster than neurons.

Since nanobots follow instructions, it is possible to imagine that nanorobots could be inserted into the body, programmed to do their job and then self-destruct the cells they have invaded. A good thing if the invaded cells are cancer cells, a bad thing, if they are good cells. Those bots would be a new kind of self-disposing poison.

3.2.4 Molecular nanosensors

Sensors at nanoscale are being developed to detect trace amounts of selected materials. These include physical (measuring mass, speed, etc.), chemical (binding strength, detecting the presence of particular materials or gasses) and biological sensors (viruses and bacteria), with great precision.[40] Such sensors will be quite useful in forensic police work since in advanced form they will be able, for example, to detect where a person has been by sampling environmental clues on clothes. They will add to the battery of information available to physicians (both human and digital) by extending and measuring minute quantities, giving them an animal's ability to detect pheromones. But one can also envision how such personal information might be used by criminals. Or, looking from another angle, this technology might enable people with malintent to develop and use "molecular camouflage" as parasitic nematodes trick their hosts by covering themselves with host molecules.[41] Can this be a potential aid for a perpetrator?

FESTOS experts assessed the likelihood of future security threats from nano-technologies. The results are shown in Figure 3.2.

Nanotechnologies pose new security threats both in the short and long term. While several nanotechnologies are maturing, using them for malicious purposes will not be easy, and most potential threats considered in the survey were expected to be of medium severity. While the

---

[40] Honey Church, K., *Nano Sensors For Chemical And Biological Applications*, Woodhead Publishing.
http://www.sciencedirect.com/science/article/pii/B9780857096609500175?np=y
[41] Perry, R., Wharton, D., (ed), *Molecular And Physiological Basis Of Nematode Survival*, CAB International, Cambridge MA, 2011.

participants judged that some technologies brought increasing threats over time, in others they judged that the threat would diminish in the long run – probably because the experts envisioned growing counter-measures availability. Molecular manufacturing seemed to lead the others, perhaps because it is perceived as the most flexible and enabling many possibilities. However, a more futuristic technology such as nanoassemblers might be realized only in the far future; its likelihood of posing a threat increases with time and its final outcome--which will be high--cannot yet be seen on the graph.

Figure 3.2 Likelihood of Nano-Technologies to pose a security threat

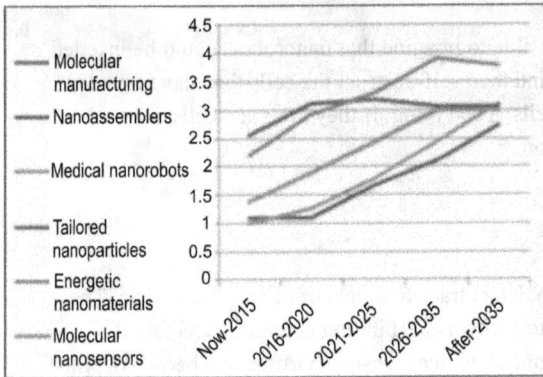

3.3 Robotics

Robotics is a rapidly developing field worldwide in civil as well as military applications. Applications that were once considered fiction have become real.

The field of robotics encompasses a broad spectrum of technologies in which computational intelligence is embedded in machines, creating systems with capabilities far exceeding the core components alone. Such robotic systems are able to carry out tasks that are unachievable by conventional machines or even by humans working with conventional tools. In particular, the ability of a machine to move and act autonomously to perform dull repetitive and dangerous tasks and even tasks that involve seemingly intelligent decisions, opens up an enormous range of applications that are uniquely suited to robotic systems.

The U.S. leads in such areas as robot navigation in outdoor environments, robot architectures (the integration of control, structure and computation), and in applications to space, defense, underwater systems and some aspects of service and personal robots. Japan and Korea lead in technology for robot mobility, humanoid robots, and some aspects of service and personal robots (including entertainment). Europe leads in mobility for structured environments, including urban transportation. Europe also has significant programs in eldercare and home service robotics. Australia leads in commercial applications of field robotics, particularly in such areas as cargo handling and mining, as well as in the theory and application of localization and navigation.

Robotic vehicles (mostly developed in the defense and space fields) are machines that move autonomously on the ground, in the air, undersea or in space. In general, such vehicles move using their own power, with sensors and computational resources onboard to guide their motion. Usually, such unmanned robotic vehicles involve some form of human oversight or supervisory control or even more direct remote control. A higher level of autonomy is an important trend of emerging technologies, and in certain future systems there will probably be zero (or almost zero) human intervention. The autonomy of robotic vehicles is heavily dependent on technological factors such as available power and energy efficiency and sociological factors such as the level of decision making allocated to the machines.. Important technological developments in this aspect include energy harvesting means (eg. solar cells, temperature gradients, currents, wind, vibrations, and biological batteries), and energy management techniques; important sociological factors include the development of adequate decision-making algorithms.

3.3.1 Biomimetics

In the evolution of robotics, there is a long tradition of utilizing biological systems as inspiration and models for new technological solutions. Several labs carry out biomimetics research for robotic applications. Long-term foresight studies show a path towards integration of humans, machines and data, and beyond lies biological integration including biotechnology and robot control utilizing means for detection of human thought.[42]

A particularly interesting emerging development in robotics is possible use of cooperative robotic teams and robotic swarms. Swarm robotics is a novel approach to the coordination of large numbers of robots, inspired by the behavior of swarms in nature, e.g. social insects that show how a large number of simple individuals can interact to create collectively intelligent systems. The EU projects like SYMBRION have developed novel principles of adaptation and evolution for symbiotic multi-robot systems, which will be able to self-configure and even self-evolve, based on bio-inspired approaches and advanced computing paradigms. In the EU project I-SWARM researchers envision that tiny (about 4 millimeters size) robots could be mass-produced in swarms and programmed for a variety of applications, such as surveillance, micro-manufacturing, medicine, cleaning and more. Now imagine that it would be possible that self-adaptation and self-reprogramming were intentionally employed for malicious behavior of the

---

[42] See, for example: Institute of Physics, "Thumbs-up for mind-controlled robotic arm," *ScienceDaily*, 16 December, 2014. http.sciencedaily.com/releases/2014/12/141216212051.htm

swarm. Such robots could be difficult to detect, difficult to anticipate, and relatively easy to deploy. Consider the uses of such swarms to distribute poisons, or to slip under doors for eavesdropping, or deployed to surreptitiously collect DNA samples or even kill.[43]

In a keynote address to the UK Royal United Services Institute (RUSI) in March 2008, Professor Noel Sharkey, a robotics and AI expert at the University of Sheffield, expressed his concerns that we are witnessing the first steps towards a robot arms race and that it may not be long before robots become a standard terrorist weapon to replace the suicide bomber. According to Sharkey, "With the current prices of robot construction falling dramatically and the availability of ready-made components for the amateur market, it wouldn't require a lot of skill to make autonomous robot weapons."[44]

3.3.2 AI-based Robot-Human Interaction and Co-existence

Japan and S. Korea, are preparing for the human-robot coexistence society which is predicted to emerge before 2030.[45] In that context researchers propose categorizing future robots as third existence entities: neither living/biological (first existence), or non-living/non-biological (second existence). A striking feature of this development is the notion of "social robots" with artificial intelligence (AI), with which people will have emotional and even intimate interactions.

These expected advances in AI and robotics are perceived as having serious potential security threats and it is not difficult to imagine what could result from an abuse of this technology by terrorists or criminals. Imagine a "Robot terrorist." Its threat will be strongly dependent on the level of artificial intelligence embedded. Imagine further that an entity that is definitely not human but more than machine begins to infiltrate our everyday lives. Regulators are trying to address the emerging safety and legal issues that are expected to emerge. No doubt that these developments will lead to new opportunities for malicious use of robots that are expected to serve us in day-to-day life.

---

[43] Zennie, M., "Death from a swarm of tiny drones: U.S. Air Force releases terrifying video of tiny flybots that can hover, stalk and even kill targets," *Daily Mail*, MailOnline, February, 19, 2013. http://www.dailymail.co.uk/news/article-2281403/U-S-Air-Force-developing-terrifying-swarms-tiny-unmanned-drones-hover-crawl-kill-targets.html

[44] University of Sheffield, "Killer Military Robots Pose Latest Threat To Humanity, Robotics Expert Warns." *ScienceDaily*, 28 February 2008. http.sciencedaily.com/releases/2008/02/080226213451.htm

[45] Weng, Yueh-Hsuan, Chen, Chien-Hsun, and Sun, Chuen-Tsai. "Toward The Human-Robot Co-Existence Society: On Safety Intelligence For Next Generation Robots", *International Journal of Social Robotics*, 1(4), pp.267-282, 2009, DOI: 10.1007/s12369-009-0019-1. http://works.bepress.com/weng_yueh_hsuan/1

3.3.3 Autonomous Mini Robots: Toys and Other Objects

Small robots with relatively high levels of autonomy are being developed for non-military applications, e.g. for medical applications as well as for the toy industry. Experts foresee that future mobile mini/micro-robots have a high potential for intrusion into privacy. They could covertly enter offices or houses--even through the crack under the door. Imagine also that such small robots would be able also to destroy or injure people in a similar way to an insect--maybe even using an insect-like body. Physicist and arms control expert J. Altman suggests limiting the use of robots below a certain size and small robotic toys in order to prevent the possibility of missuse by terrorists in public areas.[46]

3.3.4 Cyborg Insects

Engineers are controlling insects through electronics by implanting electrical stimulators that energize certain nerves or brain cells to trigger an impulse to move in a desired direction. The insects, which can be controlled by remote control or a pre-programmed chip, may soon be able to self-generate the electricity required to control them, prolonging their powered life span.[47]

A DARPA project aims to co-opt the way some insects communicate to give early warning of chemical attacks, locating disaster victims, monitoring for pollution and gas leaks, etc. Researchers have already created living communication networks by loading a package of radio transceivers, batteries, and microphones on the backs of dark-seeking cockroaches to develop a system for finding victims trapped under rubble[48]. The firm OpCoast, based in New Jersey, was awarded a contract to develop a mobile communications network using insects. The electronics package can contain an acoustic sensor designed to respond to the altered calls of other insects. This should help ensure that an "alarm" signal triggered in one cyborg insect is passed quickly across the network of nearby insects and is ultimately picked up by ground-based transceivers.[49] Each network could use hundreds or thousands of insects, though they could be spread far apart.

[46] Altman, J., *Military Nanotechnology: Potential Applications And Preventative Arms Control*, Routledge, 2006 http://books.google.ca/books?id=vebFKA3pCdQC&lpg=PR1&dq=Altman%2C%20J.%2C%20Military%20Nanotec hnology%3A%20Potential%20Applications%20And%20Preventative%20Arms%20Control%2C%20Routledge&lr &pg=PP1#v=onepage&q&f=false.

[47] Sato, H., and Maharbiz, M., "Recent Developments In The Remote Radio Control Of Insect Flights" *Frontiers In Neuroscience*, 2010; 4: 199, December 8, 2010. http://www.ncbi.nlm.nih.gov/pmc/articles/PMC3100638/

[48] McCally, K., "A new Kind of Rescue Hero" University of Rochester, http://www.rochester.edu/pr/Review/V75N1/0501_epstein.html

[49] Zyga, L.,"Cyborg Crickets Could Form Mobile Communications Network, Save Human Lives': Phys.Org, July 13, 2009: http://phys.org/news166715517.html

Researchers also envision that within a few years controlled cyborg insects could be carrying mini-cameras or other sensory devices to be used for a variety of sensitive missions including espionage.[50]

While there may be many other applications for cyber-insects that use nano and bio-technology, the use of unmodified but infected insects to spread disease has historical precedent can also be imagined: a swarm of insects that can transmit virulent agents aimed at a group of people or targeted at a crop. Crime and terror could certainly make use of such a technology.[51]

3.3.5 Robots and artificial limbs

Various applications of robots could be implemented in human bodies. For example, research is being carried out on powered leg prosthetics using electromyography signals for control, and on prosthetic hands[52]. These technical features are being developed for recovery of physical functions after accidents or diseases. While such applications are beneficial to society as a whole and paraplegics in particular, the technology could also enhance skills of persons with malintent and open new hacking pathways directly to the brains of people who use electromyography for limb control.

3.3.6 Ethical Control of Robots

The adoption of autonomous robots, particularly in military environments, leads to new problems of control. Of high concern are means for introducing ethical considerations into autonomous decision-making. The development of ethical controls has thus become a new field in computer science and philosophy. The application of autonomous systems in civilian (e.g. domestic) environments will lead to the use of such ethical control systems in new areas. One can now envision hacking that leads systems to make disastrous decisions or reconfiguring such systems for terrorist or criminal use.

Figure 3.3 shows the likelihood that the selected technologies will pose future threats, as judged by the FESTOS experts. The potential threats from all of the technologies are expected to rise with time. Autonomous robots seemed to be the most threatening technology with a sharp rise initially and holding that position over time. The likelihood of the security threat of other technologies was seen to decrease probably because experts assumed that counter measures will be found. The same assessment showed also that the main threat of robotic applications will be

---

[50] Anthes, E., "The Race to Create Insect Cyborgs." From her book Frankenstein's Cat; The Guardian, 16 Feb 2013: http://www.theguardian.com/science/2013/feb/17/race-to-create-insect-cyborgs
[51] Lockwood, J., "Six Leg And Soldiers; Using Insects As Weapons Of War", Oxford University Press. October 2008. http://www.oupcanada.com/catalog/9780195333053.html
[52] See op. cit. "Thumbs Up For Mind Controlled Robotic Arm"

to humans rather than infrastructure. The feeling among the FESTOS participants was that terrorists will be likely to utilize robotic systems for malicious purposes and cause great numbers of casualties in this way.

Figure 3.3 Likelihood of Robotic technologies to pose a security threat

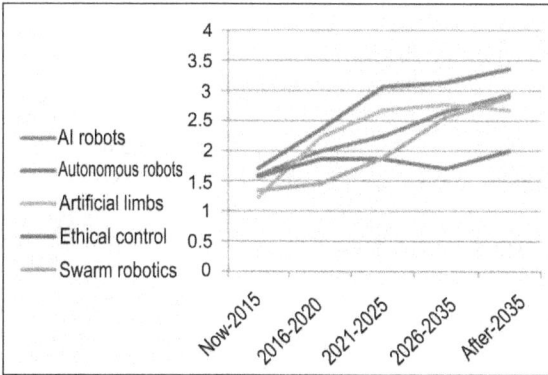

3.4 New Materials

Advances in Materials Science and Engineering have enormous impact on future manufacturing industries and new products. New technological approaches and capabilities, particularly nanotechnologies, have provided new insights into nanoscale and molecular states of materials and new ways to create, process, and eventually use them. In the last decades the materials sector has developed specialized materials for narrow niche applications and this trend is likely to continue, by utilizing high-performance alloys, nanocomposites, super-strength materials, new laminates, carbon nanotubes and other nanomaterials, a variety of special coatings, and new biological or bio-inspired materials. These are just a few examples of the richness of this field. Important areas are: smart materials and "multi-functional materials" that respond in a desired manner to changing external conditions; these are being developed for use in sophisticated structural, electronic, optical, magnetic, and biotech applications. A multifunctional material is a composite or hybrid of several distinct material phases in which each phase performs a different function. New functional materials will enable adaptable performance, mission tailor-ability and morphing: that is, adaptively changing from a material with one set of characteristics to another. With increasing complexity, multifunctional materials become not mere materials but sophisticated systems with controllable features, for example artificial muscles and self-healing, to name just two examples.

An exciting and imaginative future trend that deserves special attention is programmable matter. Wil McCarthy, an aerospace engineer and Science Fiction writer, described this vision in the magazine *Nature*:

> "4 July 2100. The flick of a switch: a wall becomes a window becomes a door. Any chair becomes a hypercomputer, any rooftop a power or waste-treatment plant. We scarcely notice; programmable matter pervades our homes, our workplaces, our vehicles and environments. There isn't a city on Earth — or Mars, for that matter — that isn't clothed in the stuff from rooftop to sub-basement ....In the 22nd century... any competent designer will simply define the shape and properties they require, including 'unnatural' traits such as super-reflectance, refraction matching (invisibility), and centuple bonds far stronger than diamond, then distribute the configuration file to any interested users." [53]

He sees quantum dots (nanocrystal particles with unique features) as a possible building block of matter:

> "The unique trait of a quantum dot, as opposed to any other electronic component, is that the electrons trapped in it will arrange themselves as though they were part of an atom, even though there's no atomic nucleus for them to surround. Which atom they emulate depends on the number of electrons and the exact geometry of the wells that confine them, and in fact where a normal atom is spherical, such 'designer atoms' can be turned into cubes, tetrahedra or any other shape, and filled with vastly more electrons than any real nucleus could support, to produce 'atoms' with properties that simply don't occur in nature" [54].

DARPA is not waiting till the 22nd century, and is already funding Programmable Matter research to create materials that can be programmed to self-assemble, alter their shape to perform a desired function, and then disassemble. Among the expected breakthroughs in this program are encoding information into chemistry, or fusing materials with machines, and fabrication of mesoscale particles with arbitrary complex shapes, composition, and function. Known also as 'InfoChemistry', this emerging field combines chemistry, information theory, and programmability to build information directly into materials. [55]

Dr. M. R. Zakin, program manager for Programmable Matter project at DARPA, described a science-fiction-like scenario to illustrate their expectations:

---

[53] McCarthy, W., "Programmable Matter", *Nature 407, 569*, October 5, 2000. http://www.nature.com/nature/journal/v407/n6804/full/407569a0.html

[54] Ibid.

[55] See for example: Samuel W. Thomas III et al.," Infochemistry And Infofuses For The Chemical Storage And Transmission Of Coded Information", Proceedings of The National Academy of Sciences of The United States of America, March 10, 2009. http://www.pnas.org/content/106/23/9147

"In the future a soldier will have something that looks like a paint can in the back of his vehicle. The can is filled with particles of varying sizes, shapes and capabilities. These individual bits can be small computers, ceramics, biological systems—potentially anything the user wants them to be. The soldier needs a wrench of a specific size. He broadcasts a message to the container, which causes the particles to automatically form the wrench. After the wrench has been used, the soldier realizes that he needs a hammer. He puts the wrench back into the can where it disassembles itself back into its components and re-forms into a hammer.... You're blurring the distinction between materials and machines. Materials act like computers and communications systems, and communications systems and computers act like materials."[56]

Such technological developments, if achieved, will have tremendous impact on our lives. The DARPA project is aimed at defense applications, but the research (carried out in leading universities like Harvard, MIT and Cornell) has high potential for important civil applications ranging from aerospace to medicine. Morphing materials can be used to change an aircraft's wings in flight or in clothing that alters its characteristics to keep users cool in the day and warm at night. One of the possible future directions is programming adaptability into the material itself. Such adaptability, for example, could produce electronic devices that can adapt to heat and dust in the desert and then shift to resist humidity and moisture in a jungle environment.

But new opportunities are also opened to terror. Future materials with enhanced controllable features may be attractive to terrorists in various ways, from enhancing explosive properties to sophisticated concealment, camouflage, invisibility, protection of covert activities, development of new toxic chemicals and more. Terrorist groups have already shown their capability to develop and manufacture super-toxic lethal chemicals and highly contagious biological agents. This is partially enabled by open access to scientific and technological developments that can be used in criminal attacks.

3.4.1 Metamaterials and Optical Cloaking

Metamaterials are specially engineered materials that have unusual optical properties that could enable optical "cloaking" and the creation of 'super-lenses' with a spatial resolution below that of the wavelength of visible light (e.g. below 200 nanometers). Several novel scientific discoveries in optics, materials, and nanotechnology in the last decade have enabled research groups to develop theoretical models of cloaking devices made out of "metamaterials" that can hide objects from sight, or make them appear as other objects. These are materials with a negative refractive index.

---

[56] Kenyon, S H., "Programmable Matter Research Solidifies" *Signal*, June 2009. http://www.afcea.org/content/?q=programmable-matter-research-solidifies

In 2006, a U.S.-British team of scientists successfully tested an "invisibility cloak" for microwave frequencies in the laboratory. The team, lead by David Schurig and John Pendry described a cloaking device: a copper cylinder that was hidden inside a cloak consisting of artificially structured metamaterials. The team then showed that the copper cylinder and the cloak were actually invisible to microwave frequencies for which the cloak was designed. Their results provide an example of an electromagnetic cloaking mechanism, and demonstrate the feasibility of such technologies.[57] We can now imagine that invisibility might be a practical possibility. Military applications are clear enough, but consider how appealing they could be to future terrorists.

3.4.2 Programmable Matter

These are materials that can be programmed to self-assemble or alter their shape and physical properties to perform a desired function, and then disassemble in response to user input or autonomous sensing. As mentioned earlier in this section, DARPA has funded active research programs in InfoChemistry but the field goes well beyond this. At MIT and Harvard, engineers have made origami robots that can fold into a working machine from flat laser cut pieces. There are programs that demonstrate an assembly of interlocking parts that are 10 times stiffer than ultra light materials of the same weight. Smart sand is another technology in which the grains pass messages to each other in order to assemble into a whole sculpture.[58]

The potential for terrorists are apparent. One can imagine the use of easily reconfigurable tools with perfect performance including weapons (that can pass security checks at airports) and that can be ready and adaptable to changing conditions and requirements.

3.4.3 Future fuels and structural materials for nuclear technologies

New materials and processes for safe and efficient nuclear reactors have been developed, such as organic superconductors, materials with special magnetoresistive effects, radiation-induced segregation, uranium silicide fuels that require only low-enrichment, new solid lubricants, nanocrystalline diamond films, etc. These lead to better understanding of the mechanisms of irradiation-induced swelling of materials, improved prediction of the behavior of fuel elements in reactor cores, development of inelastic neutron scattering techniques, determination of properties of trans uranium compounds, and more. Such developments may be in the more distant future

---

[57] Schurig, D., et al. "Metalmaterial Electromagnetic Cloak at Microwave Frequencies," *Science*, November 10, 2006. http://www.sciencemag.org/content/314/5801/977.short
[58] MIT News, Programmable Matter: http://newsoffice.mit.edu/topic/programmable-matter

but during R&D, opportunities for theft of radioactive or toxic materials will increase and these advanced materials might find their way into "dirty" bombs or other weapons.

3.4.4 Water catalyzing explosive reactions

Research has shown that water, in hot and dense environments, can play an unexpected role in catalyzing complex explosive reactions. Imagine a terrorist IED that requires only water to initiate a reaction. These devices could be completed at the location at which they are to be exploded. They could construct bombs in place, in real time and ignite them by simply adding water.

3.4.5 Personal rapid prototyping and 3D printing

Current 3D printers can construct products after downloading a CAD/CAM description from a computer or scanning an object to be copied by the printer. The lowest cost printer in 2015 was about $1,000 and manufacturers were speculating about lowering that cost by a factor of 10.[59] Printers are able to make any 3D plastic model that one can think of--from flower pots, self portrait statues and Lego bricks, to mechanical parts, motherboards and vehicle replacement parts. The threat stems from the potential ability of LWs (or indeed any criminal) to construct weapons of plastic from CAD/CAM plans available in digital form or from scanning other weapons.

Personal rapid prototyping machines can use several different raw materials: thermoplastics (Polylactic acid), Ceramic slurries (Silicon nitride), Silicone polymer, Wood's metal, Field's metal, and even Chocolate. Further planned is the introduction of non-polymeric materials like certain metals, ceramics and composites. The opportunity opened to crime and terrorism is remarkable. Personal rapid prototyping machines can potentially be used for creating weapons. Polymers are widely used in guns, and criminals can design new types of cheap guns that only need several simple metal parts, or combine common metal parts with specially designed plastic parts. Future generations of personal rapid prototyping machines will be able to work with hard ceramic materials that will open other new opportunities for weapons. Polymer weapons may also pass metal detectors without been detected. Besides guns, other plastic weapons can be manufactured as well. Commander Mark Simmons, Head of Scotland Yard anti-knife project, said:

> "We have had some finds of non-metallic knives. The numbers are not vast at the moment, and they do not start me thinking that there is a major new trend. We have seen some made out of carbon fiber, a type of hard plastic. A few of those have been found.

---

[59] The Peachy Printer - The First $100 3D Printer & Scanner! https://www.kickstarter.com/projects/117421627/the-peachy-printer-the-first-100-3d-printer-and-sc

People will be ingenious about this sort of thing, and there will always be individuals thinking of different ways around a new policing tactic."[60]

All in all, the development envisioned in the hardware on the one hand and the variety of materials to be used on the other hand will open a huge market for home made weapons of many kinds. The idea of self designed and built guns or new types of weapons on the streets is disturbing: availability of weapons will be easier and their control and detection more difficult.

FESTOS experts assessed the likelihood of threats posed by several different material technologies. Their judgments are shown in Figure 3.4. We see that clearly 3D printing was viewed as the most threatening technology in the long run. The other technologies seemed less threatening, perhaps because they are more difficult to be realized. However their possible use should also be taken into account.

Figure 3.4 Likelihood of Materials technologies to pose a security threat

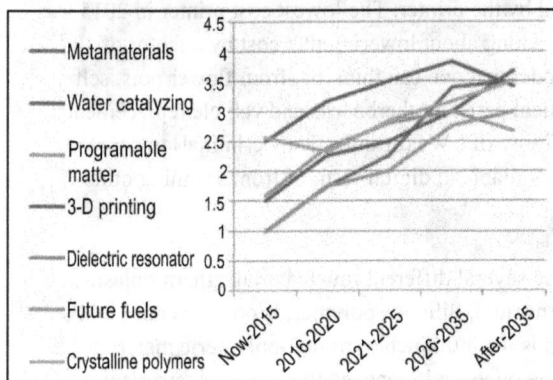

3.5 Ease of Implementation

The potential of a technology to serve malintent purposes differs. Some of these technologies are easier for individuals to apply than others; some will have more severe consequences than others when applied. In FESTOS, participating experts examined ease and severity and defined a third parameter--the potential of malicious use--as the product of the two. First, with respect to ease, note that a full development process is not required. A terrorist can make use of a technology without completing all safety protocols that are sometimes difficult to perform. For the second parameter, the level of severity of a potential threat, FESTOS experts considered the impact on

---

[60] Barrett, D., "Plastic Knives Used to Evade Metal Detectors," *The Telegraph*, December 6, 2008
http://www.telegraph.co.uk/news/uknews/law-and-order/3628285/Plastic-knives-used-to-evade-metal-detectors.html

both people and infrastructure. The ratings of the two parameters--easiness and severity--were made on a scale of 1 to 5.

Table 3.1 shows the average results derived from all experts in column A and B. Column C shows the multiplication of A and B.

Table 3.1 The potential for malicious use of a technology[61]

Technology	A: How easy will it be to use this technology for malicious purposes?	B: How severe is the potential security threat posed by this technology?	C: Potential for malicious use of the technology
1. New gene transfer technologies	3.52	3.22	11.33
2. Synthetic biology	3.16	3.40	10.74
3. Cyborg insects	3.33	3.08	10.26
4. Energetic nanomaterials	3.00	3.33	9.99
5. Autonomous & semi-autonomous mini robots	3.36	2.83	9.51
6. AI-based robot-Human Interaction	3.00	2.94	8.82
7. Swarm robotics	2.89	3.00	8.67
8. Water catalyzing explosive reactions	2.56	3.38	8.65
9. Self-replicating Nanoassemblers	2.75	2.92	8.03
10. Personal rapid prototyping and 3D printing machines	2.89	2.71	7.83
11. Metamaterials with negative light refraction	2.50	2.95	7.37
12. Tailored nanoparticles	2.53	2.89	7.31
13. Future fuels	2.33	3.07	7.16
14. DNA protein interaction	2.58	2.58	6.65
15. Programmable matter	2.29	2.79	6.39
16. Molecular manufacturing	2.50	2.50	6.25
17. Medical nanorobots	2.27	2.73	6.20

---

[61] Source: Hauptman, A., Raban, Y., Katz, O., Sharan, Y., "Foresight of Evolving Security Threats Posed by Emerging Technologies," WP2: Horizon scanning, page 132 : http://www.sicherheitsforschung-europa.de/servlet/is/14805/FESTOS%20Final%20report%20on%20potentially%20threatening%20technologies.pdf?command=downloadContent&filename=FESTOS%20Final%20report%20on%20potentially%20threatening%20technologies.pdf

Technologies in this table are arranged according to their potential for malicious use. Gene transfer, Synthetic Biology and Cyborg insects are highest in the list. They received a relative high easiness rating as well as a high severity rating. Other technologies from robotics and nanotechnology followed.

Judging from these results, LWs will have a growing menu of technologies from which to select and apply to weapons of choice. Many applications will be cheap, effective, and easy to acquire.

Terrorist capabilities will come not only from the well-known NBC (Nuclear, Biological, and Chemical) fields but also from robotics and nanotechnology. All can give rise to new weapons with massive effects. Even more, technologies like 3D printing will enable a potential terrorist to produce weapons covertly at home using materials available on the market. These will make the SIMAD scenario more likely. It seems inevitable that technological progress will bring with it new opportunities for terrorists to increase their deadly potential.

While we have concentrated on the possible acquisition and uses of advanced technologies by LWs, existing toxins are available to them as well, and these are commercially available, deadly, stockpiled, with known recipes for production, and hard to trace. These Toxic Industrial Chemicals (TIC) include chlorine, phosgene, arsine, anhydrous ammonia, pesticides, and others. The chemical accident in Bhopal India in 1984 resulted in a toxic cloud that killed 2,500 people immediately and 16,000 after months or years.[62] To underscore the point that informed terrorists, including lone wolves, can manufacture their own weapons, remember the Aum Shinrikyo cult in Japan which began their attempts to create a theocratic state by attacking the town of Matsumoto in June 1994 with sarin gas, and the Japanese subway system in 1995. They sought to manufacture 70 tons of sarin nerve gas in a factory they purchased from Switzerland. Sarin is a terrible, internationally outlawed weapon, yet it has been used by nations against their people (Syria, Iraq); so, stockpiles probably exist despite international agreements. These and other stocks of CBW may seem like fair game to lone wolves who have aspirations to appear in history books and plan to exist in eternal bliss in their heaven.

3.6 Counteractions

The potential for terrorist use of advanced technologies calls for the development of anticipatory measures to prevent their use in attacks and to mitigate their effects. Many new technologies are designed to help humanity overcome current and futures challenges; however lack of awareness about possible mal-use might also lead to abuse of the same technologies. With forethought, the

---

[62] Tucker, J., "The Future of Chemical Weapons," *The New Atlantis*, Fall 2009.
http://www.thenewatlantis.com/publications/the-future-of-chemical-weapons

mal-use threat can be reduced. In the following section we describe some policy principles and activities that may be possible in this direction.

There is a long history of attempts to reach worldwide agreement on biological and chemical weapons production, use, stockpiling, and destruction. The Chemical Weapons Convention (CWC) bans chemical weapons and requires that existing weapons be destroyed over a given time period. As of October 2013, the CWC had 190 parties.[63] Israel and Myanmar have not yet ratified the convention, while Angola, Egypt, North Korea, and South Sudan have neither signed nor acceded to the convention. Destruction of chemical weapons as required by the convention is proceeding and is verified by the Organization for the Prohibition of Chemical Weapons.

Biological weapons fall under the Biological Weapons Convention (BWC); as of April 2015, it had 173 State parties and 9 signatories.[64] However, verification of compliance to this convention is difficult because biological production programs can have dual purposes: for military applications or for veterinary, medical, and agricultural uses. Inspectors could file with OPCW but their capabilities are very limited. Instead, parties are to file complaints about violations with the UN Security Council. While chemical weapons are being destroyed under CWC, the bioweapon stockpiles remain.

Smallpox, as a biological weapon in these stockpiles, has been classified as a Class A bioterrorism agent by the Center for Disease Control; Class A agents are easy to distribute, can be transmitted from person to person and thus may initiate an epidemic, and are deadly. (Other Class A agents are: Anthrax, Tularemia, Plague, Botulism, and Viral hemorrhagic fever.[65]) Smallpox has been responsible for epidemics around the world for thousands of years (estimated 300 million deaths in the 20th century.[66]). In a medical triumph that has been called the most important medical advance of the twentieth century, the disease was declared eradicated in 1980.[67] The last known case was in Somalia in 1977.[68] Prophylactic vaccinations essentially stopped and because of low prevalence, natural immunity diminished.

The World Health Organization recommended that all countries destroy remaining samples of the virus and as far as is publicly known, the United States and Russia retained the only two

---

[63] CWC Status of Participation, OPCW http://www.opcw.org/about-opcw/member-states/status-of-participation/
[64] Membership of the Biological Weapons Convention, UNOG
http://www.unog.ch/80256EE600585943/(httpPages)/7BE6CBBEA0477B52C12571860035FD5C
[65] Centers for Disease Control and Prevention, Bioterrorism Agents/ Diseases;
http://emergency.cdc.gov/agent/agentlist-category.asp
[66] Flight, C., "Smallpox: Eradicating the Scourge," *BBC*, Feb 2, 2011:
http://www.bbc.co.uk/history/british/empire_seapower/smallpox_01.shtml
[67] Global alert and response: smallpox world health organization: http://www.who.int/csr/disease/smallpox/en/
[68] Smallpox Surveillance: Worldwide, CDC, Oct 24, 1997:
http://www.cdc.gov/mmwr/preview/mmwrhtml/00049694.htm

remaining samples of the virus. These samples were closely guarded. In 1996, WHO recommended that these samples also be destroyed[69] but on the advice of scientists in the field, they were retained to aid in research should the disease return and for use in other research.

Yet the threat of weaponized smallpox and other bioweapons has persisted. Stockpiles of biological and chemical weapons exist--some ostensibly in the Middle East; where wars flare, the safeguards must be in question. One wonders how safe the safeguards really are. In July 2014, the U.S. Centers for Disease Control announced that more unrecorded vials of live virus had been discovered by chance in a cardboard box in a refrigerator at the National Institute of Health in Washington.[70]

Safeguards are needed to deny LW opportunities for theft of critical materials especially if they are part of a relevant lab or research establishment. More than that; conventions like CWC and BWC are not adapted to deal with individuals or even with non-state actors. It should also be mentioned that the use of technologies like synthetic biology can result in agents that are not listed in the formal conventions and are thus not controlled by them. This is especially true when chemicals are concerned. Inclusion of more and more materials in such conventions is not easy and we might find in the future new materials that are not included in such control regimes. This trend is even more worrisome because present conventions are not dealing with other fields of technology that might be the source of more new "non-conventional" threats. Fields like nanotechnology and robotics could potentially give rise to such capabilities as well as materials that actually bypass present conventions. This situation calls for serious consideration of new conventions that will help decrease the evolving new threat. However, fighting against new threats needs more than effective conventions.

Other counteractions are needed. On the one hand, we need good passive protection against emerging threats. On the other hand, we should develop good proactive ways to stop the threats and decrease their potential impacts. These include early identification of potential LWs and effective policies to prevent them from achieving their purposes. In addition, we should plan development of identification methods of possible new threatening materials as well as means against to protect against them.

Sensors that detect and warn of dangerous conditions are, of course, already in wide use. Nuclear radiation detectors, toxic gas detectors, explosive sniffers, oxygen depletion sensors, indicators of the presence of selected bacteria, and virus identifiers: all are used by industry, police, firefighters, security agencies, medical laboratories, and the military. In countering terrorist

---

[69] Altman, L., "Final Stock Of The Smallpox Virus Now Nearer To Extinction In Labs", *New York Times*, January 25, 1996
[70] Reardon, S., "NIH Finds Forgotten Smallpox Store", *Nature*, http://www.nature.com/news/nih-finds-forgotten-smallpox-store-1.15526

attacks, sensors of all types need to be rapid in identifying the threat, sensitive to very small quantities, of broad enough scope to catch threats that were unanticipated, and accurate enough to minimize false positives and negatives. New future technologies promise to offer some of these characteristics.

Real time detection using gel-based microchips is proving attractive. In this approach, arrays of antibodies developed from the toxins in animals are labeled with fluorescent dyes, exposed to the suspect atmosphere, and sampled through fluorescence, laser-based mass spectroscopy or some other technique to detect which antibodies are activated and thus the presence of dangerous materials. The sensitivity is excellent and the time required for analysis is very short.[71] Scientists and engineers at Lawrence Livermore National Laboratories are developing a "microbial detection array," a three-inch chip with over 350,000 probes that can identify over 2,000 viruses and 900 bacteria.[72]

At MIT's Lincoln Laboratories, systems are being developed to detect the presence in the atmosphere of aerosols with signatures that reveal whether or not they differ from the background environment of natural aerosols using a process called laser-induced fluorescence. Automated analysis allows sensitive discrimination between threatening and non-threatening particles.[73]

Another approach to early detection of a bio-attack is known as syndromic surveillance. In this technique, statistical analyses are continually performed on public health records to identify as early as possible deviations from expected health conditions. If new cases of what seems to be the flu are seen, the question can be "is the spread within statistical expectations or is it an induced phenomenon?"[74] This approach has a number of problems associated with it such as excessive delay time and similarity of symptoms between "standard" diseases such as flu and terrorist-induced diseases with flu-like symptoms. However, the advent of big databases and automation of medical records could certainly improve the speed, sensitivity and accuracy of this technique.

Another action designed to limit the potential threat of unintended applications of gene synthesis was taken in 2008 by a European consortium of companies called the International Association

---

[71] Rubina, A. Yu, et al., "Quantitative Immunoassay Of Biotoxins On Hydrogel-Based Protein Microchips," *Analytical Biochemistry*, May 2005. http://www.sciencedirect.com/science/article/pii/S0003269705000862

[72] Dillow, C., "3 Inch Bio Detector Quickly Scans For All The Bacteria And Viruses We Know, All At Once," *Popular Science*, May 8, 2010. http://www.popsci.com/science/article/2010-05/three-inch-bio-detector-scans-all-bacteria-and-viruses-we-know-all-once

[73] Primmerman, C., "Detection of Biological Agents," *Lincoln Lab Journal*, v12, Number 1, 2000. https://www.ll.mit.edu/publications/journal/pdf/vol12_no1/12_1detectbioagents.pdf

[74] *Syndromic Surveillance: An Effective Tool For Detecting Bioterrorism?* Rand Corporation 2004. http://www.rand.org/content/dam/rand/pubs/research_briefs/RB9042/RB9042.pdf

of Synthetic Biology (IASB). They initiated the formulation of a code of conduct that includes special gene-screening standards. Formulation of codes of conduct has continued ever since.[75]

However, experts still worry that these activities may be insufficient. There is an increasing threat of wide proliferation of new genetic engineering technologies and equipment in many (non-military) biological laboratories that could be used for malicious and illegal activities. Reproductive cloning raises well-known ethical concerns--and also concerns about illegal activity. Even novel techniques developed for disease immunization might be abused and applied for precisely the reverse effect.

In order to address these concerns we need well-established, multi-agency R&D policies that would take such problems into account in national and international planning. The objectives of such planning would be to develop various measures that increase protection of people against unintended uses of novel materials and agents. Programs like these could include:

- Accelerating R&D of methods of detection, identification, therapy, and mitigation of adverse effects of largely unknown technologies in development including materials, bio-materials, and nanoparticles of various types.

- Developing a system for monitoring and limiting the use of molecular manufacturing devices that are available to the public.

- Developing effective ways to control instrumentation and knowledge dissemination, particularly in the areas of bioengineering and synthetic biology, which pose threats of unintended use in creating cheaper and widely accessible tools for bioweapons.

- Strengthening research towards wide spectrum medicines to counter various biological agents (antiviral and antibacterial drugs).

- Performing foresight studies on potential threats of "DNA - Proteins interaction" technologies in criminal/terrorist hands.

- Developing methods for detection of non-radio opaque weapons.

- Requiring developers of new technologies to anticipate the potential terror uses of the technologies they are developing and to research in parallel, detection and mitigation methods (an impact statement?).

- Assigning higher priority to threats with high likelihood or high potential for malicious use.

- Inculcating a sense of responsibility for both intended and unintended consequences of new technologies and applications among scientists, engineers, and policy makers.

---

[75] Codes of conduct exist in many disciplines, particularly disciplines in which errors could be catastrophic and for which dual use is apparent. See for example: http://projects.exeter.ac.uk/codesofconduct/Chronology/

An additional action that should be considered is adoption of concepts like "security by design" when systems are built. According to this concept, developers have to think about the potential of abuses of future products and develop means to avoid such possibilities. This principle should be applied in robotics in order to limit the abuses that robots might create and to limit their use for malicious purposes. This would include development of specific built-in safeguards for robotic systems as well as measures to limit possible hacking into robotic systems.

Finally, comprehensive legal and regulatory safeguards should be developed to help assure that R&D activity and the resulting applications consider, from the outset, the potential for abuses. You will find information on this aspect further in this book.

3.7 Conclusions

Science and Technology are developing fast and opening new opportunities and applications for people with malicious plans. Fields like nanotechnology, robotics, new materials and biotechnology are leading in this vicious process. The future terrorist will be more technologically-skilled and will use those skills to reach for weapons and other tools at the leading edge of development. Weapons will be smaller, easier to hide, sometimes undetectable, more efficient, and cheaper. Terrorists could produce them at "home" covertly or even at the site of use and apply them when and where they consider appropriate. Many of these new weapons could cause large-scale casualties making the attacks significant and dramatic. SIMADs are plausible. A variety of conventional and non-conventional means will be available to them and their presence is likely to be intensely felt. Past events and experience show that this assessment is not a speculation, but a sound forecast.

4. The Cyber Dimension

4.1 Introduction

The scope and spectrum of cyber-security is continuously expanding, addressing a wider and more diverse range of "enemies", from the "basement geek" to criminal organizations and nation-state actors. Cyberspace has become a new medium for competing commercial, ideological, or political adversaries, and the main battlefield of law enforcement organizations against extremists and cybercriminals. In the U.S., annual breaches have increased roughly five-fold over the past five years[76], and the seriousness of digital disruptions is also rising dangerously.

The Internet can play several important roles in the evolution of LW terrorism: a passive role in, for example recruiting LWs and educating them in the ways of the lone wolf. It can also play an active role in bringing about large-scale destruction, for example in causing vital services such as the international flow of funds to crash. Third, Internet and the media in general communicate information about the success or failure of LW operations and thereby glorify the LW, fuel or deter copycats, and excite public interest and attention. Fourth, the Internet also can be used in a forensic mode to trace messages on the Internet back to their origins and perhaps gain early warning about LW terrorist intent.

The Internet is the proverbial two-edged sword: terrorists can use the Internet for espionage or sabotage, to accomplish disruptive cyber-attacks, recruit and proselytize, raise money, and create a PR image, while the security and national agencies can use Internet defensively to guard against attacks, to root out the LWs and SIMADs before they have a chance to do damage or to capture them after the fact, and in some cases, to offensively attack institutions and systems of their enemies.

Intelligent hackers can act alone, drawing on information commonly available, without attracting much attention. They can set up conventional web sites for dissemination of information and fundraising, or they can put on their "black hats" and execute their cyber-plans without threat of physical harm to themselves, to achieve destruction and create chaos. They can steal information for resale, as was the case in the theft of data from Anthem Insurance, the second largest health insurer in the U.S. In this theft the hackers collected millions of records of customers and staff

---

[76] Lisa Monaco, Assistant to the President for Homeland Security and Counterterrorism; Cyber Threats and Vulnerabilities: Securing America's Most Important Assets, Woodrow Wilson Center, February 10, 2015 http://www.wilsoncenter.org/event/cyber-threats-and-vulnerabilities-securing-americas-most-important-assets

including names and addresses and Social Security numbers. Anthem apparently had not encoded the data.[77]

Hackers can attack for political reasons, as was apparently the case in the data theft from Sony Pictures; in this case the hacker was thought to be from North Korea.[78]

How do they gain access? They may have or gain administrator's credentials, which make entry simple. Many operating systems and hardware devices are though to have "back doors" that allow monitoring and in some cases, hackers may enter through these portals. "Trojan horses" provide the means to insert false information or for assuming system control. Sometimes it's not about the money, but for the thrill and to demonstrate prowess and skill to impress their peers.

Perhaps the most famous case of cyber crime is the theft attributed to Edward Snowden, a computer professional and whistleblower, of an estimated million documents[79] of classified information from NSA's computer files pertaining to "military capabilities, operations, tactics, techniques and procedures"[80] as well as the agency's surveillance programs and spying devices, beginning in June 2013. Because Snowden demonstrated that NSA was privy to essentially all communications and digital transactions, the level of distrust grew after his disclosures began: security systems were revised, international hackles were raised, apologies exchanged, foreign countries became suspicious of U.S. companies operating within their borders "as witting or unwitting collaborators in the U.S. government's surveillance and intelligence gathering activities."[81] Some people call Snowden a whistleblower hero, others, a traitor.

Herve Falciani is another whistleblower who, unlike Snowden, was apparently concerned with data on money laundering and tax cheats. He is thought to have stolen large amounts of data (100,000 name files) from his employer, HSBC in 2007 and then turned the set over to French authorities. These data included account information--names of the people involved, the amount of money in the accounts, and revealing notes about the client needs for secrecy: money laundering, tax evasion. The money records totaled billions of euros.[82] Without identifying

---

[77] Toor, A., "Anthem failed to encrypt customer data prior to cyberattack," *The Verge*, Feb 6, 2015; http://www.theverge.com/2015/2/6/7991283/anthem-hack-encrypted-data
[78] Sony Hack. Continuing coverage of the hacking of Sony Pictures Entertainment. http://www.nbcnews.com/storyline/sony-hack
[79] Estimate by Keith Alexander, retired chief of NSA: http://en.wikipedia.org/wiki/Edward_Snowden#Global_surveillance_disclosures
[80] From testimony by Army Gen. Martin Dempsey: http://www.nbcnews.com/news/investigations/snowden-leaks-could-cost-military-billions-pentagon-n46426
[81] Scott Kennedy, Director of the Research Center for Chinese Politics and Business at Indiana University: http://www.reuters.com/article/2014/01/21/us-ibm-china-idUSBREA0K0FB20140121
[82] Brinded, L., "Switzerland Charges Ex-HSBC Whistleblower Herve Falciani With Industrial Espionage," *Industrial Business Times*, December 11, 2014. http://www.ibtimes.co.uk/switzerland-charges-ex-hsbc-whistleblower-herve-falciani-industrial-espionage-1479122

Falciani by name, the office of the Swiss Attorney said it was proceeding to try him in absentia for industrial espionage:

> "The Swiss Attorney General's office, without identifying the suspect as is its custom, said in a statement today that the country was prepared to try him in absentia. The statement refers to Herve Falciani, the technician accused of stealing client data in 2008 from HSBC's Geneva office and passing it to French authorities, said a person with direct knowledge of the case who asked not to be identified.
>
> HSBC was charged in France last month with money laundering through tax fraud and illegal marketing in a case stemming from the data theft. The bank faces a similar investigation in Belgium and last month agreed to pay $12.5 million to settle claims with the Securities and Exchange Commission that its Swiss private-banking unit solicited U.S. investors without being registered....
>
> In its statement, the Attorney General's office said the defendant has sometimes been 'celebrated as a hero abroad'."[83]

Did he do it for anti-crime purposes or for profit? Following is an excerpt from the interview conducted by TV news magazine "60 Minutes":

> "Bill Whitaker (Interviewer): *Did you want to be paid for this?*
>
> Hervé Falciani: *I want it, of course. I deserve that.*
>
> Bill Whitaker: *You deserve to be paid?*
>
> Hervé Falciani: *But I knew that in Europe it was impossible.*
>
> Bill Whitaker: *It does leave one wondering if you did this for high-minded purpose, or because, as you were saying, you thought there was profit in this?*
>
> Hervé Falciani: *Yeah, common profits. I have no problem with profit. I knew that at the same time, I could have both.*"[84]

Another whistle-blower, Everett Stern, was assigned to the HBSC anti-money laundering department and reportedly uncovered significant information that showed that despite warnings from regulators, HBSC was involved with money laundering and worse. He is alleged to have found that:

> "There was an exchange company wiring large sums of money to untraceable destinations in the Middle East. A Saudi fruit company was selling millions, Stern found with a simple Internet search, to a high ranking figure in Yemeni wing of the Muslim brotherhood. Stern even learned that HSBC was allowing millions of dollars to be moved

---

[83] Miller, H., "Ex-HSBC Employee Falciani Said Indicted on Data Theft," *Bloomberg Business*, December 11, 2014. http://www.bloomberg.com/news/articles/2014-12-11/ex-hsbc-employee-falciani-said-indicted-on-data-theft
[84] 60 Minutes, CBC, February 8, 2015: http://www.cbsnews.com/news/hsbc-swiss-leaks-investigation-60-minutes/

from the Karaiba chain of supermarkets in Africa to a firm called Tajco, run by the Tajideen brothers who had been singled out by the treasury department as major financiers of Hezbollah."[85]

According to reports, when his bosses were unresponsive, he left the company and took his information to the FBI.

Joseph Konopka, also known as Doctor Chaos, was one of a number of infamous, lone wolf cyber-terrorists. In middle school, according to his mother, he was the "go to" person for fixing computer problems. He became bored and dropped out of school and recruited others who wanted to use computers and information technology to cause mayhem; this group was known as "The Realm of Chaos."[86] He and his group shut down power companies and did other damage on-line for thrills and to demonstrate their capabilities. According to FBI records:

"Milwaukee agents identified another "lone-wolf" terrorist whose actions nearly escalated to the use of chemical weapons. From 1998 through 2002, Joseph Daniel Konopka, also known as 'Dr. Chaos,' was "indicted by a Federal Grand Jury in Milwaukee on 13 counts covering 53 Wisconsin crimes. From 1998 through 2002, Konopka wreaked havoc in 13 counties by disrupting power and causing $800,000 in damages. He was also accused of setting fires, disrupting radio and television broadcasts, disabling an air traffic control system, selling counterfeit software, and damaging the computer system of an Internet service provider. Konopka was arrested in March 2002 after being caught with cyanide, a potentially deadly chemical, near the Chicago subway system. ...He was later sentenced to 13 years in prison."[87] [A portion of this sentence was later overturned.[88]]

4.2. Education, fund raising, and recruitment

The features of the Internet and social media that make them attractive for so many users also make them attractive for terrorist recruitment, propaganda, fund raising, and education: they are essentially instantaneous, worldwide, accessible through inexpensive computers and hand-held devices, and provide a level of anonymity to users. Internet carries both "official" and unofficial websites of terrorist organizations, as well as "distributor" sites that give links to sources of

[85] Taibbi, M., "Gangster Bankers, Too Big to Fail," *Rolling Stone*, February 13, 2014.
http://www.rollingstone.com/politics/news/gangster-bankers-too-big-to-jail-20130214

[86] Hamill, D. S., and Heinzmann, D., "Computer Whiz to Dr. Chaos," *Chicago Tribune*, March 13, 2002
http://articles.chicagotribune.com/2002-03-13/news/0203130362_1_computer-sabotage-joseph-konopka

[87] FBI, A Brief History, Milwaukee Division: http://www.fbi.gov/milwaukee/about-us/history-1

[88] "'Dr. Chaos' arson conviction overturned," *Associated Press*, June 1, 2005.
http://www.nbcnews.com/id/8055451/ns/us_news-crime_and_courts/t/dr-chaos-arson-conviction-overturned/

information and materials of possible use. Social media encourage networking among likeminded proto-jihadists and their teachers. According to the *Jerusalem Post*:

"Information technology has enabled terrorist organizations to receive and share knowledge globally. Terrorists can easily obtain information on sensitive targets and their potential weaknesses; public transport timetables; building sites, their opening times and their layout. Al-Qaida maintains an extensive database that contains information about potential American targets."[89]

Based on the comments of some RTD participants, do-it-yourself weapons, including biological weapons, are likely to be considered or used by LW terrorists. The Internet and social media provide ideal communication tools for spreading information about these techniques.

Many commentators have described the use of terrorist videos showing horrendous executions, immolations, beheadings, and war action as material designed to stimulate recruitment, including encouragement of single actors. One wonders why a contrary marketing campaign can't be mounted against such an obvious and disgusting ploy. After all, the U.S. has a masterful capacity for marketing. Why not campaigns aimed at the appropriate demographic with slogans that say, "Don't fall for that rot" or "Don't be a jerk" or "They must think you're stupid"? Or an episode of the Simpsons that shows Bart deluded by a terrorist web site[90], or memorable lyrics that evoke the spirit of anti-terror heroes?

4.3 Active destruction via Internet

A U.S. Department of Homeland Security paper *The Internet As A Terrorist Tool For Recruitment And Radicalization Of Youth*, reported that "In 1998 there were a total of 12 terrorist-related websites active. By 2003, there were approximately 2,630 sites, and by January 2009 a total of 6,940 active terrorist related websites."[91] Other sources have estimated that 15,000 sites existed in July 2009.[92] Although data are hard to find, there is every indication that since this report was issued in 2009, the number of sites and their sophistication could have grown considerably.

---

[89] "In Internet's Way: Terrorism on the Free Highway," Jerusalem Post, Oct 23, 2013
[90] One episode of the *Simpsons* dealt with suspected terrorism, however. It was titled "Mypods and Broomsticks," Episode 7 of season 20, aired originally on November 30, 2008.
[91] US Department of Homeland Security, Washington DC, citing a presentation given by Dr. Gabriel Weimann at the Youth Recruitment and Radicalization Roundtable, March 19, 2009.
[92] "American-Bred Terrorists Causing Alarm For Law Enforcement," *ABC News*, July 22, 2010
http://abcnews.go.com/WN/suspected-american-terrorists-islamic-ties-causing-concern-law/story?id=11230885&page=1

Even before the Internet, in 1991, when the pre-Internet was still ARPANET, a self replicating virus was written by Bob Thomas of Bolt, Beranek and Newman, to infect DEC computers and demonstrate the danger of self replicating computer viruses. It led to the display "I'm the creeper, catch me if you can."[93] A series of malware viruses and worms followed with names like Michelangelo and SQL Slammer, generally developed by young people with the purpose of demonstrating their digital prowess, until in 2010 Stuxnet appeared, a purpose-designed worm that was targeted specifically at Siemens controllers that were running nuclear-enrichment centrifuges in Iran's nuclear program. The worm caused the centrifuges to over-speed and destroy themselves. The author is not yet known and Iran has been quiet on the whole affair.

Some relevant comments from the participants in the RTD study include:

Cyberweapons to knock out power grids or damage critical infrastructure presently exist.

Cyber terrorism is the most likely vehicle for successful, sustained attack by a lone wolf.

If cyber terror is included (it ought to be) then the answer (to the question when might we expect a large scale cyber-attack) is now.

Consider including Cyber Terrorism as a WMD. Cyber terror has been defined as: the intimidation of civilian enterprise through the use of high technology to bring about political, religious, or ideological aims, actions that result in disabling or deleting critical infrastructure data or information and/or resulting in massive loss of life.

Like these members of the RTD panel, the former U.S. Secretary of Defense, Leon Panetta, was also deeply concerned about the possibility of a massive cyber attack against the U.S.:
"The most destructive scenarios involve cyber actors launching several attacks on our critical infrastructure at once, … could also seek to disable or degrade critical military systems and communications networks. The collective result of these kinds of attacks could be "cyber Pearl Harbor": an attack that would cause physical destruction and loss of life, paralyze and shock the nation, and create a profound new sense of vulnerability."[94]

The attacks that the Defense Secretary is talking about could come from nation states (as Stuxnet might have), terrorist groups or even single LW hackers; after all, many act-alone hackers derive notoriety among their peers from designing and launching complex viruses and worms.

---

[93] Kushner, D., "The Real Story of Stuxnet: How Kaspersky Lab tracked down the malware that stymied Iran's nuclear-fuel enrichment program," *IEEE Spectrum*, February, 2013. http://spectrum.ieee.org/telecom/security/the-real-story-of-stuxnet

[94] Leon Panetta, Defense U.S. Secretary Speech, *Defense News*, Oct. 12, 2012. http://www.defensenews.com/article/20121012/DEFREG02/310120001/Text-Speech-by-Defense-U-S-Secretary-Leon-Panetta

Imagine an advanced technology such as AI or nanobots in the hands of a potential SIMAD. Nick Bostrom presents several doomsday scenarios, none are very pleasant but many are germane to the SIMAD issue. One of his scenarios envisions self-replicating nanobots:

> "Such (self replicating, bacterium scale) replicators could eat up the biosphere or destroy it by other means such as by poisoning it, burning it, or blocking out sunlight. A person of malicious intent in possession of this technology might cause the extinction of intelligent life on Earth by releasing such nanobots into the environment."[95]

That's an example of a SIMAD.

A recent report by The Millennium Project contained a scenario that read:

> "After land, sea, air, and space, cyberspace became the "fifth battlespace" on the agenda of security experts. Governments and businesses are under cyberattacks daily (espionage or sabotage) from other governments, competitors, hackers, and organized crime. The sources of these assaults are difficult to verify, making retribution difficult, and when sources are verified, what are the appropriate responses? Countries, especially highly connected ones, have to consider the threat of a cyber-Pearl-Harbor; much effort is being devoted to cyber-defense and potential countermeasures. Because society's vital systems increasingly depend on the Internet, cyberweapons to bring them down can be thought of as weapons of mass destruction."[96]

Akami Technologies Inc. tracks Internet traffic and reports on the number and nature of cyber attacks. During the third quarter of 2014, Akamai found that attacks were originating in 201 countries and regions, 49% of the attacks originating in China. [97] The trend today is for many of these attacks to be in the form of Distributed Denial of Services (DDoS). The Syrian Electronic Army claimed credit for some of these attacks, by compromising domain name registrations to divert traffic away from intended recipients.

The Internet-of-Things (IoT) will offer LW hackers new and rich targets. The IoT refers to the interconnection of devices, automated command systems, and sensors designed to control or diagnose the machinery of a city, country, business, or household, to detect their presence, location, and proper functioning. Hacking into the systems by which these devices communicate and their data are stored offers an enticing target: by 2020 as many as 75-80 billion devices may be connected through the Internet. [98] In fact such an attack occurred in December 2013. According to the security firm Proofpoint, hackers penetrated home appliances over a two week

---

[95] Bostrom, N., "Existential Risks. Analyzing Human Extinction Scenarios and Related Hazards", *Journal of Evolution and Technology*, Vol. 9, No. 1 (2002). http://www.nickbostrom.com/existential/risks.html
[96] Op cit. *State of the Future*
[97] The State of the Internet, 3rd Quarter, 2013, The Akamai Corporation, http://www.akamai.com/dl/akamai/akamai-soti-q313.pdf?WT.mc_id=soti_Q313
[98] op.cit. *2013-14 State of the Future*

period invading "home-networking routers, connected multimedia centers, televisions and at least one refrigerator to create a platform to deliver malicious spam or phishing emails." [99] The TV news magazine *60 Minutes* demonstrated on camera, how a hacker could take over control of an automobile (which, after all, carries many computers) disabling the brakes of a target vehicle, for example.[100]

4.4 Encryption and Communications

The open-source OpenSSL cryptography library contains a set of algorithms that programmers can use to encrypt their computer code. However the codes for encryption contained bugs that could be exploited by hackers. One in particular, the HeartBleed security bug, provided a means for hackers who took advantage of the bug to eavesdrop on conversations, and steal passwords and other data thought to be secure. Many other bugs have been discovered in the library, but the HeartBleed bug was easy to exploit and left no trace when used.[101] It was disclosed on April 7, 2014. By then, some half a million (about 17%) of the Internet's certified secure web servers were believed to be vulnerable. This means that any information that was transmitted on those servers was potentially readable by anyone equipped with enough knowledge—including cyber-criminals. "Some might argue that it is the worst vulnerability found (at least in terms of its potential impact) since commercial traffic began to flow on the Internet,"[102] wrote Joseph Steinberg, in *Forbes*.

The Israeli cyber intelligence biweekly report describes recent attacks via Internet. Here's one example:

> "The cyber consulting company, Cyber Hat, reported the spread of the Cryptolocker malware in some Israeli companies, which is a ransomware deployed to a company network by phishing and encrypts files onto the company servers. The original email from Cryptolocker comes with a ransom note demanding payment of 400 Euros through Bitcoin currency in return for the decrypting files."[103]

Anders Breivik had 7,000 Facebook friends and he chose to publish his 1,500-page manifesto there just hours before his attack in Norway killed 93 people, encouraging readers to consider his

---

[99] Fox News: "Hackers Use Refrigerator in Cyber Attack," January 20, 2014. http://www.foxnews.com/tech/2014/01/20/hackers-use-refrigerator-in-cyber-attack/

[100] 60 Minutes, February 8, 2015

[101] The Heartbleed Bug http://heartbleed.com/

[102] Steinberg, J., "Massive Internet Security Vulnerability -- Here's What You Need To Do", *Forbes*, April 10, 2014. http://www.forbes.com/sites/josephsteinberg/2014/04/10/massive-internet-security-vulnerability-you-are-at-risk-what-you-need-to-do/

[103] "Two weeks of diverse cyber-attacks in Israeli cyberspace", Executive Cyber Intelligence Bi-Weekly Report by INSS-CSFI, March 19, 2014, http://i-hls.com/2014/03/executive-cyber-intelligence-bi-weekly-report-inss-csfi-4/

manifesto a plan for action.[104][105] In commenting on his manifesto, Bakker and deGraff say that this book can be:

> "…regarded as a guide into the workings of lone operator terrorism. In one part of his manifesto, Breivik also points at the possibilities of the use of unconventional weapons, such as Radiological Dispersal Devices, or so-called dirty bombs… Breivik explains how to publish documents on the Internet and how to use social media for recruiting purposes. Moreover, he shows the tricks he himself used to circumvent European custom agents and describes in detail how he manufactured the explosives he used to blow up the government building in Oslo."[106]

While Breivik--and before him, Kaczynski and others--wanted to reach the largest possible audiences with their manifestos, some LWs want to keep at least a portion of their pre-event communications secret and private; use of encryption is commonplace. A recent UN Office on Drugs and Crime report notes:

> "… software programs are available to disguise or encrypt data transmitted over the Internet for illicit purposes. These programs may include the use of software such as Camouflage to mask information through steganography[107] or the encryption and password protection of files using software such as WinZip. Multiple layers of data protection may also be employed. For example, Camouflage allows one to hide files by scrambling them and then attaching them to the end of a cover file of one's choice. The cover file retains its original properties but is used as a carrier to store or transmit the hidden file. This software may be applied to a broad range of file types. The hidden file may, however, be detected by an examination of raw file data, which would show the existence of the appended hidden file."[108]

Many data-masking or data-obfuscation techniques exist, but at some level the underlying data must always be available and is therefore vulnerable.

---

[104] Halford, M., "Anders Breivik. Reader," The New Yorker, July 24, 2011.
http://www.newyorker.com/online/blogs/books/2011/07/anders-breivik-manifesto.html
[105] Boston, W., "Killer's Manifesto: The Politics Behind the Norway Slaughter", *Time World*, July 24, 2011.
http://content.time.com/time/world/article/0,8599,2084901,00.html
[106] Bakker, E., and de Graaf, B., "Preventing Lone Wolf Terrorism: some CT Approaches", *Perspectives on Terrorism*, Vol5, Issues 5-6., December 2011
op cit. Bakker and deGraff
[107] Secret writing or concealing messages within images or other messages.
[108] "The Use Of The Internet For Terrorist Purposes," UN Office on Drugs and Crime, New York, NY, September 2012. This quotation includes a reference from UNDOC: "Written submission of expert from the Raggruppamento Operativo Speciale of the Carabinieri of Italy."

4.5 Forensic, Offensive, and Defensive Uses of IT

The three most well-known offensive uses of IT may well be the Stuxnet virus, the attack on Sony Pictures, and Wikileaks/NSA surveillance, but many other offensive uses and technologies are candidates for the list.

An IP address is associated with all Internet communications and generally, can be used to identify a region of the world from which a message emanates and with authorization, an Internet Service Provider can pinpoint the particular subscriber account being used. For example, traffic flow into jihadist websites can be monitored to identify frequent users and, in turn, their email and social network communications can be traced. So, while LW terrorists operate independently, their sources of information and communications often can be used to identify influential sources and links to others. However, increasingly smart programmers can successfully camouflage the real IP addresses.

In answer to the question "What technology is likely to be most effective for the detection of people with evil intentions?", one respondent to the RTD said:

> Precisely what NSA is currently accused of--eavesdropping (spying), on *everyone* is what it will take. Despite this activity running contrary to the American vision/expectation, it is the most effective means of learning what someone is thinking about, planning to do.

The National Security Agency does more than monitor communications. Their Tailored Access Operation (TAO) has claimed that they have developed and used an extensive array of software and firmware techniques for interception of messages, identifying who said what, when, and where, and implanting false information. Will these activities which raise so many questions about privacy and civil liberties help stop LWs and SIMADs? The answer is not yet clear.

In December, 2013, *The Washington Post* described a portion of the agency's system of collecting mobile phone location records, which apparently yields about five billion records a day.[109] According to the *Washington Post*, "the records are stored in a huge database known as FASCIA, which received over 27 terabytes of location data within seven months."[110]

In December 2013, the German magazine *Spiegel* published a catalog of devices and software that ostensibly have been developed and used by the NSA in its efforts to discover and thwart

---

[109] Gellman, B., and Soltani, A., "NSA tracking cellphone locations worldwide, Snowden documents show". *The Washington Post*, December 4, 2013. A *Washington Post* video about this program can be viewed at: http://www.washingtonpost.com/world/national-security/nsa-tracking-cellphone-locations-worldwide-snowden-documents-show/2013/12/04/5492873a-5cf2-11e3-bc56-c6ca94801fac_story.html
[110] "Ghostmachine: The NSA's cloud analytics platform" *The Washington Post*, Retrieved December 28, 2013 (from a Wikipedia bibliography in an article titled: "Global surveillance disclosures (2013–present)")

terrorist activities. [111] [112] Most of these were labeled "Top Secret" but contained prices indicating that they may also have been intended for use by others. The listing of devices and software (complete with prices which are generally fairly low, considering the technology involved), developed by NSA is long and impressive, indeed. Two examples from this list are eavesdropping and locating devices:

DROPOUTJEEP is a software implant for the Apple iPhone that utilizes modular mission applications to provide specific SIGINT functionality. This functionality includes the ability to remotely push pull files from the device. SMS retrieval, contact list retrieval, voicemail, geo location, hot mic, camera capture cell tower location, etc. Command control and data exfiltration can occur over SMS messaging or a GPRS data connection. All communications with the implant will be covert and encrypted.[113]

TAWDRYYARD is used as a beacon, typically to assist in locating and identifying deployed RANGEMASTER units. Current design allows it to be detected and located quite easily within the 50 foot radius of the radar system being used to illuminate it... the simplicity of the design allows the form factor to be tailored for specific operational requirements. Future capabilities being considered are return of GPS coordinates and a unique target identifier and automatic processing to scan the target area.[114]

It is possible to buy malware systems commercially that have capability similar to those being developed by TOA at NSA. According to on line sources[115], Lench IT solutions PLC sells a sophisticated commercial surveillance toolkit (FinSpy, FinFisher) that includes "location tracking, remotely activating a built-in microphone and conducting live surveillance via "silent calls," as well as the ability to monitor all forms of communication on the device, including emails and voice calls."[116] A comprehensive global Internet scan by the Citizen Lab of the Munk School of Global Affairs at the University of Toronto found command and control servers of FinFisher's surveillance software as part of Gamma International's FinFisher "remote monitoring solution," in 25 countries: Australia, Bahrain, Bangladesh, Brunei, Canada, Czech Republic, Estonia, Ethiopia, Germany, India, Indonesia, Japan, Latvia, Malaysia, Mexico,

---

[111] "NSA's ANT Division Catalog of Exploits for Nearly Every Major Software/Hardware/Firmware", *Spiegel*, December 29, 2013, through leaksource.info: http://leaksource.info/2013/12/30/nsas-ant-division-catalog-of-exploits-for-nearly-every-major-software-hardware-firmware/

[112] "NSA's top secret technology can tap/infect computers even when not connected to the Internet", *Kurzweil AI Weekly Newsletter*, January, 15, 2014. http://www.kurzweilai.net/nsas-top-secret-technology-can-tapinfect-computers-even-when-not-connected-to-the-internet

[113] DROPOUTJEEP https://leaksource.files.wordpress.com/2013/12/nsa-ant-dropoutjeep.jpg

[114] TAWDRYYARD https://leaksource.files.wordpress.com/2013/12/nsa-ant-tawdryyard.jpg

[115] FinFisher, FinSpy, from Wikipedia: http://en.wikipedia.org/wiki/FinFisher

[116] Schwartz, M., "FinFisher Mobile Spyware Tracking Political Activists", Information Week, August 31, 2012. http://www.darkreading.com/vulnerabilities-and-threats/finfisher-mobile-spyware-tracking-political-activists/d/d-id/1106086?

Mongolia, Netherlands, Qatar, Serbia, Singapore, Turkmenistan, United Arab Emirates, United Kingdom, United States, Vietnam.[117]

Other software systems that automate hacking for the hackers are available on line. Until its software was leaked, the ZeuS Crimeware kit sold for about $10,000 mostly to potential hackers who wanted to use the system to gain access to legitimate computers and servers for a variety of reasons that may have included stealing information for resale, monitoring conversations to obtain competitive information or for blackmail.[118] The price has dropped because Zeus itself apparently has been hacked.[119] Its uses are well known:

> "Computers, smart phones and tablets infested with the Zeus bot (zbot) malware become agents for criminals—serving a malicious master, sharing user data, and becoming part of a botnet to attack computer systems. Using the kit, attackers could harvest data, such as usernames and passwords, as entered in a Web browser on an infected device. In addition, an attacker may insert additional fields into the display of a Web form on a legitimate Web site to trick the user into supplying more data than a site usually requires, such as a PIN number on a banking site. Attackers can even remotely request the user's machine take a screenshot of the current display at any time. All data requested by the attacker is sent back to a command and control panel, where it can be sorted, searched, used, or sold. The harvested data is likely to be used for identify theft but could also be sold to a company's competitors or used to publicly embarrass a firm."[120]

Developers of software try to identify weakness in their programs and fix them before hackers can discover and exploit them. Hackers also try to find these weaknesses, if possible before the developers do. When hackers attack through previously unknown holes they have an opportunity to mount a "Zero-Day" attack. The challenge for cyber security experts is to produce a defensive system that can guard against all unknown (but not necessarily unknowable) attacks before they occur, obviously a tough assignment.

As for the future, Dan Kaufman, who is in charge of developing software innovation at DARPA, says DARPA is "working on artificial intelligence software that would detect a hacker attack in real-time and plug it in milliseconds with no humans involved. If such technology had been available to Sony, that breach allegedly from North Korea could have been plugged right as it

---

[117] Marquis-Boire, M., Marczak, B., Guarnieri, C., and Scott-Railton, J., *You Only Click Twice: FinFisher's Global Proliferation*, March 13, 2013. https://citizenlab.org/2013/03/you-only-click-twice-finfishers-global-proliferation-2/

[118] Goodin, D., "Source code leaked for pricey ZeuS crimeware Kit", *The Register*, May 10, 2011. http://www.theregister.co.uk/2011/05/10/zeus_crimeware_kit_leaked/

[119] Coogan, P., "Zeus, King of the Underground Crimeware Kits", *Symantec Connect*, 25 Aug 2009: http://www.symantec.com/connect/blogs/zeus-king-underground-crimeware-toolkits

[120] "Akamai Warns Fortune 500 of High-Risk Threat from Zeus Crimeware", June 10, 2014. Akamai, http://www.akamai.com/html/about/press/releases/2014/press-061014.html

---

happened."[121] DARPA is also developing new search strategies that will penetrate the darknet and maintain associative relationships known as Memex. On a *60 Minutes* interview, Chris White, to whom the invention of Memex is attributed, is using the system to identify networks of connections in its first application: sex trafficking. Memex "can comb through all the sex ads online--over 60 million--and identify 100s of names and numbers that link together and make up an entire trafficking ring". The Manhattan District Attorney told *60 Minutes* that the city had 20 open investigations in which they were using Memex tools to track down sex offenders.[122]

4.6 Conclusions

This information battle-space has all of the elements of a real war. There are offensive attacks for ideology, profit, or revenge, there is intelligence gathering to blunt the offense, there are defenses that can be erected and overcome with the right weapons. Analogous to snipers there are single person attacks, and analogous to armies there are battalions of hackers that can attack in unison. There are big weapons such as those developed by NSA, and small weapons (but potentially powerful) of the amateur LW hacker. R&D is important as the information weapons escalate in power. There are unintended casualties of large-scale attacks such as loss of identity of people when the real target may have been a government. In the U.S., several cyber-defense units, have been created including a cyber command within the Department of Defense,[123] a Cyber Response Group (CRG) at the Homeland Security Department, and a new Cyber Threat Intelligence Integration Center under the auspices of the Director of National Intelligence to provide and rapidly integrate information of different counter-terrorism agencies.[124]

Despite cyber-security efforts:

- Nuclear plant computer systems have been hacked (but not the sacrosanct control elements as far as we know).[125] The hacker or group responsible for the current cyber

---

[121] "DARPA: Nobody's safe on the Internet" *60 Minutes*, February 8, 2015. http://www.cbsnews.com/news/darpa-dan-kaufman-internet-security-60-minutes/

[122] Ibid.

[123] "The U.S. Army Cyber Command's breadth of responsibility spans the entire Army and the entire world - from the tactical edge to the strategic enterprise-level or national levels. Traditional boundaries no longer exist and anonymous attacks can occur literally at near light speed over fiber optic networks. Our enemies will deny the freedom of movement on our networks and use whatever they can from wherever in the world they are to gain advantage." U.S. Army Cyber Command, Army Cyber: http://www.arcyber.army.mil/org-arcyber.html

[124] Monaco, L., Assistant to the President for Homeland Security and Counterterrorism, *Cyber Threats and Vulnerabilities: Securing America's Most Important Assets*, Woodrow Wilson Center, February 10, 2015. http://www.wilsoncenter.org/event/cyber-threats-and-vulnerabilities-securing-americas-most-important-assets

[125] Durden, T., *Nuclear Power Plant In South Korea Hacked*, December 21, 2013. http://www.zerohedge.com/news/2014-12-23/nuclear-power-plant-south-korea-hacked

---

penetration in South Korea posted blueprints of the nuclear reactors and threatened to release more internal information.

- Massive data thefts have occurred including millions of names or files, for example, recently, from Anthem, Target, and Sony, NSA, and others for sale, blackmail, embarrassment, or political purposes.

- Data has been stolen to provide competitive advantage at a national level. A news release of the U.S. Department of Justice describes an indictment of five Chinese military hackers by a grand jury in Pennsylvania; they are accused of "computer hacking, economic espionage and other offenses directed at six American victims in the U.S. nuclear power, metals and solar products industries. The indictment alleges that the defendants conspired to hack into American entities, to maintain unauthorized access to their computers and to steal information from those entities that would be useful to their competitors in China.[126]

- Financial fraud via Internet has been so extensive that the U.S. Secret Service has created an Electronic Crimes Task Force (the Nigerian "send money" fraud, credit card fraud, money laundering, and electronic counterfeiting fall in this category.)

As important as all of these challenges are, perhaps the most significant and underreported is the insertion of false information into otherwise valid data storage systems. As an illustration, Wikipedia keeps track of fraudulent entries that are detected. Over the years fictional people with elaborate biographies have been invented, along with synagogues, a made up language that is described as about to become extinct, non existent musical instruments, opening chess moves, and where Nicholas Copernicus is said to have contracted genital warts from a Minsk prostitute, to name a few from a much longer list.[127]

As amusing as this list is, it is borne on the winds of great concern. If false and manipulated information can be inserted in Internet systems, then, inevitably, our ability to rely on it for valid data will be eroded, possibly seriously. At the extreme, consider an international financial system or even just the stock market functioning when the data on which it is based has been shown to be corrupt. And what of the future when human minds are directly connected with information libraries on the Internet or when mind to mind transfer of information, emotion, experience, and memories are possible: would these be hacking targets? How would they fare in an era of falsified information?

---

[126] US Department of Justice, "U.S. Charges Five Chinese Military Hackers for Cyber Espionage Against U.S. Corporations and a Labor Organization for Commercial Advantage", *Justice News*, May 19, 2014. http://www.justice.gov/opa/pr/us-charges-five-chinese-military-hackers-cyber-espionage-against-us-corporations-and-labor

[127] List of hoaxes on Wikipedia: http://en.wikipedia.org/wiki/Wikipedia:List_of_hoaxes_on_Wikipedia

---

5. Detection of Potential Lone Wolves

Can potentially destructive persons be detected before they act to become LW or SIMAD? The early identification of persons who plan to do harm may be the best counteraction of all.

5.1 Introduction

In this chapter, we discuss some current and future strategies for identifying potential lone wolf and SIMAD terrorists before they have a chance to act. In a sense, this is "pre-crime" police work and is a field that will grow in importance if, as we suspect, lone wolves gain access to massively destructive power. The pre-detection idea appeared in a science fiction short story by Phillip K. Dick, *The Minority Report*, in 1956, followed by a film adaptation in 2002.[128] In that story, the principal means of anticipation was the clairvoyance of some "precog mutants" who had visions of crimes to come and sent the police to intercept the "would be" criminals before they had a chance to act. In this chapter we explore more solid technologies that may have a chance to be used to reduce the probability of terrorist crimes.

Techniques for identifying potential terrorists who are linked to extremist groups are becoming increasingly successful, but in the case of LWs, it is very difficult--perhaps impossible--to sense when an individual becomes a threat to society. One reason is that LWs are an extremely diverse group: some, like the convicted Fort Hood Shooter, Major Nidal Malik Hasan, are disciplined members of the society (in this specific case, even the Armed forces); others such as mathematician Ted Kaczynski, the Unabomber, are outstanding intellectuals. Some LWs are professionals, for example, Baruch Goldstein was a physician who killed and injured 150 people praying in the Cave of the Patriarchs in Hebron. Others such as Adam Lanza, the murderer of Sandy Hook Elementary School are recent students themselves. Many are described as loners, but others are not. Motivations also vary widely. Yigal Amir who was sentenced to life imprisonment for assassinating Yitzhak Rabin, Prime Minister of Israel--clearly a political motivation; but LWs also include those who kill for what they perceive as apolitical reasons, such as Scott Roeder who was convicted of killing a physician who performed abortions. As Bakker and de Graff note:

> "Lone wolfs, by definition, are idiosyncratic. They display a variety of backgrounds with a wide spectrum of ideologies and motivations: from Islamists to right wing extremists, and from confused suicidal psychopaths to dedicated and mentally healthy persons."[129]

---

[128] Dick, Philip K. (2002), "Minority Report", In: Minority Report, London, 1-43
[129] op cit. Bakker and deGraff

Therefore, it is no wonder that acts of LW terror are so difficult to anticipate.

Yet some techniques seem to be working. Since 9/11/2001, the Heritage Foundation has been tracking published information about terror plots that have been foiled. By April 2012, they had 50 such events on their list. Of these 50 examples, 24 were LWs. These cases involved LWs of various ages, with different motivations, with plans to use different weapons, and with different life experiences.[130] Here are a few examples:

> Jose Pimentel, a naturalized U.S. citizen pleaded guilty to a charge of terrorism. He was accused of planning to use pipe bombs to attack targets in New York City. [131]

> Rezwan Ferdaus was arrested in September 2011and pleaded guilty to attempting to damage and destroy a federal building; he was accused of planning an attack on the Pentagon and Capitol Building using a small drone aircraft laden with explosives.[132]

> Sami Osmakac was found guilty for attempting to use a weapon of mass destruction and possession of fully automatic firearms. He apparently planned to attack targets in the Tampa Bay area.[133]

> Amine El Khalifi was sentenced to 30 years in prison for plotting to attack the US Capitol building with guns and a bomb.[134]

We reviewed these cases to find the techniques that appear to have been most effective in identifying and apprehending these people before they could act. The methods have been:
- Receiving information from informants.
- Interdiction of by-standers (underwear bomber on board an international flight and suspicious street vendors in the foiled Times Square bombing attempt.)
- Online surveillance of chat rooms and postings to web sites on how to make WMD and solicitation to violence and in some instances even posts about the attacks they are contemplating.

---

[130] Carafano, J. J., Ph.D., Bucci, P. S., Ph.D. and Zuckerman, J., *Fifty Terror Plots Foiled Since 9/11: The Homegrown Threat and the Long War on Terrorism*, The Heritage Foundation, April 25, 2012. http://www.heritage.org/research/reports/2012/04/fifty-terror-plots-foiled-since-9-11-the-homegrown-threat-and-the-long-war-on-terrorism

[131] Jacobs, S., "Lone Wolf Terrorist Jose Pimentel gets 176 Years in Prison", *New York Daily News*, March 25, 2014. http://www.nydailynews.com/news/crime/lone-wolf-terrorist-jose-pimentel-16-years-prison-article-1.1734472

[132] Bidgood, J., "Massachusetts Man Get 17 Years In Terrorist Plot", *New York Times*, November 2 2012. http://www.nytimes.com/2012/11/02/us/rezwan-ferdaus-of-massachusetts-gets-17-years-in-terrorist-plot.html

[133] Phillips, A., "Sami Osmakac Gets 40 Years in Prison for Plotting Terrorist Attacks in Tampa", *Tampa Bay Times*, November 5, 2014: http://www.tampabay.com/news/courts/criminal/sami-osmakac-gets-40-years-in-prison-for-plotting-terrorist-attacks-in/2205214

[134] FBI: "Virginia man sentenced to 30 years in prison for plot to carry out suicide bomb attack on US Capitol", http://www.fbi.gov/washingtondc/press-releases/2012/virginia-man-sentenced-to-30-years-in-prison-for-plot-to-carry-out-suicide-bomb-attack-on-u.s.-capitol

---

- Results of a physical police search (in 2011, the backpack of a person arrested near Arlington National Cemetery was found to contain ammonium nitrate, spray paint, and spent ammunition rounds.)
- Sting operations in which undercover law enforcement officers appeared to support the LW's plot and supplied inert weapons
- Tracking associates of known terrorists
- Tracking purchases of large quantities of bomb making chemicals (e.g. phenol)

These are tried and true police techniques that have been upgraded over the years with modern technology such as DNA analysis and monitoring of social media. In the years immediately ahead, new techniques and technologies will improve the success ratio of these venerable techniques, and methods never available before will come on the scene and the game will change. The evolution of existing techniques and the appearance of new techniques and methods are the subjects of this chapter.

5.2 Automated screening of big databases

Massive amounts of data are being collected and stored by government and private organizations in pursuit of their goals and objectives. The huge repositories of health records, school records, purchase records, arrest records and the like could be the raw material of a new kind of empirical science in which cause and effect are linked, not by equations, but by data, correlating variables in one database with changes in others using powerful statistical analyses routines, to produce identification where users thought their appearance was hidden.

The size of some of the databases is already staggering, and nevertheless their number and growth are accelerating. For example:

Weather and climate data collected by the climate simulation center at NASA's Goddard's Space Flight Center stores about 32 pentabytes of data. [135]

eBay's computers store 90 pentabytes of data that record information about 100 million users and 500 million auction listings in 50,000 categories.[136]

In 2010, Facebook stored 50 billion photos supplied by its 500 million users who create 100 billion hits per day. The company stores 2 trillion objects and handles hundreds of millions requests per second.[137]

---

[135] Climate Change Simulation: NASA's Weather Supercomputer:
http://www.csc.com/cscworld/publications/81769/81773-supercomputing_the_climate_nasa_s_big_data_mission
[136] Inside eBay's 90PB Data Warehouse, http://www.itnews.com.au/News/342615,inside-ebay8217s-90pb-data-warehouse.aspx

The Utah Data Center is a data storage facility for the intelligence community of the United States. When it is completed it will house the world's largest data storage and encryption facility; data storage capacity has not yet been published but is estimated as a yottabytes (10^{24} bytes) of data. This size is necessary since essentially all communications and related data collected minute by minute by the NSA will be stored here.[138]

As big as these (and other) databases are, much more is coming. For example, the Internet-of-Things envisions the connection of household and industrial machines to each other and to sources of information that will allow them to diagnose and self correct maintenance issues, order required support and service, and exchange and record information automatically. Today some automobiles can ask, periodically, "Would you like to run a system health report?" and if one answers, "yes" it proceeds to check all systems and report on their operational status. The IoT is a network of everyday objects (food items, home appliances, clothing, etc), as well as various sensors, addressable and controllable via the Internet. It contains operating manuals for people who forget how to turn on their food blender, or better yet turns it on based on the response to a voice command.

Even old school, dumb products will be able to join in when micro-chip radio frequency identification chips are added to their labels. Manufacturers will be able to collect information about how people use their products and difficulties they encounter in the field; quality of products can be improved by observing failure modes under realistic conditions. Furthermore the Internet of things can minimize the probability of breakdown and its consequence to users--an incipient malfunction detected in a refrigerator could trigger an automated maintenance call and a parts replacement order. Perhaps these automated interactions will be covered under warranty agreements made at time of purchase.

The milieu created by the IoT is known as *Ambient Intelligence*. In a world of Ambient Intelligence, the devices on the IoT seem to think and endeavor to be helpful: advertising is adaptive, on the spot product recommendations come from talking mirrors and furniture, mobile connectedness is total, and the doorbell recalls the favorite drink of the person who has just come into your condo. The price for all of this, of course, is loss of privacy. All of the data about your things and how you use or do not use them will be in a database somewhere, and as in all databases, will be subject to compromise, theft, and distortion.

The world of Ambient Intelligence could be a hacker's paradise. Hacking into the systems by which these devices communicate and their data are stored offers an enticing target.

---

[137] Johnson, R., "Scaling Facebook to 500 Million Users and Beyond", July 21, 2010.
https://www.facebook.com/notes/facebook-engineering/scaling-facebook-to-500-million-users-and-beyond/409881258919
[138] Bamford, J., "The NSA is Building the Country's Biggest Spy Center ," *Wired*, March 15, 2012.
http://www.wired.com/2012/03/ff_nsadatacenter/all/1

Big data is expanding in other dimensions as well. Consider:

> The number of still digital photos is enormous and is growing at a staggering rate. By 2009, there were an estimated 2.5 billion camera phones in use (1/3 of the world's population). Expansion of the market to 1 billion took only 7 years from a standing start (by comparison, TV took 65 years [139]). How many photos do these phones take? On his blog, Jonathan Good reasons as follows: "If the average person snaps 150 photos this year that would be a staggering 375 billion photos. That might sound implausible but this year people will upload over 70 billion photos to Facebook, suggesting around 20% of all photos this year will end up there. Already Facebook's photo collection has a staggering 140 billion photos, that's over 10,000 times larger than the Library of Congress." [140]

And consider that video information in databases exceeds that of still photos. YouTube says that

- More than 1 billion unique users visit YouTube each month
- Over 6 billion hours of video are watched each month on YouTube—that's almost an hour for every person on Earth
- 100 hours of video are uploaded to YouTube every minute
- 80% of YouTube traffic comes from outside the U.S. [141]

In the world of Ambient Intelligence: when a "secure" database of information about you (say your credit card purchases) is combined with another "secure" data base that records entirely different information (say health records,) all security can disappear as inferences about who is being described by a different coded numbers used in each database become recognized as the same individual. In other words, attempts to anonymize data can be overcome when data in different databases are meshed. In one study "… 3 months of credit card records for 1.1 million people (showed) that four spatiotemporal points were enough to uniquely re-identify 90% of individuals…" [142]

Why is all of this important to the detection of would-be lone wolves? First of all, the ability to identify specific persons from still and video images is improving greatly; if the technology were perfect then a name could be attached to every image in the still and video databases. The Labeled Faces in the Wild (LFW) database is a standard set of photographs of faces against

---

[139] Ahonen, T., "Communities Dominate Brands, Business and Marketing Challenges for the 21st Century" http://communities-dominate.blogs.com/brands/2009/11/celebrating-30-years-of-mobile-phones-thank-you-ntt-of-japan.html

[140] Good, J., "How Many Photos Have Ever Been Taken? http://blog.1000memories.com/94-number-of-photos-ever-taken-digital-and-analog-in-shoebox

[141] YouTube Statistics, www.youtube.com/yt/press/statistics.html

[142] De Montjoye, Yves-Alexandre, Radaelli, L., Singh, V K., Pentland, A., "Unique in the shopping mall: On the reidentifiability of credit card metadata," *Science*, January 30, 2015; http://www.sciencemag.org/content/347/6221/536

---

which face recognition algorithms are tested.[143] By 2010, the level of correct identification reached about 72%; newer algorithms increased the success rate to over 90% by 2011.[144] The developers of this analysis method say:

> "Face.com has been used by users and developers to index almost 31 billion face images of over 100,000,000 individuals. Leveraging this immense volume of data presents both a unique opportunity and an unusual challenge. The capability developed in house in order to make use of this data builds upon various achievements in scientific computation, database management and machine learning techniques. The run-time engine itself is a real-time one, able to process face detection together with recognition of over 30 frames per second on a single Intel 8-core server machine."[145]

Facebook already has a significant facial recognition capability and is developing advanced versions, as reported in *Science*.

> "Facebook's DeepFace system is now as accurate as a human being at a few constrained facial recognition tasks. The intention is not to invade the privacy of Facebook's more than 1.3 billion active users, insists Yann LeCun, a computer scientist at New York University in New York City who directs Facebook's artificial intelligence research, but rather to protect it. Once DeepFace identifies your face in one of the 400 million new photos that users upload every day, "you will get an alert from Facebook telling you that you appear in the picture," he explains. "You can then choose to blur out your face from the picture to protect your privacy... One benchmark for facial recognition is identifying whether faces in two photographs from the LFW data set belong to the same celebrity. Humans get it right about 98% of the time. The DeepFace team reported an accuracy of 97.35%—a full 27% better than the rest of the field. ...."[146]

These systems in fact make all the vast store of photos into mug shots and can in principle, provide a detailed time line of who is doing what, where, and to whom.

Data mining by government agencies of massive databases raises the significant question of privacy. Do governments, particularly the U.S. government, have the right to examine personal data: phone calls, numbers dialed, travel, purchases, video files, and the like? Legal scholars argue both sides of the issue, but a national security law brief argues:

> "... the Fourth Amendment does not protect anything a person knowingly exposes to the public. The Supreme Court has upheld the legality of governmental use of pen registers

---

[143] Labeled Faces in the Wild, http://vis-www.cs.umass.edu/lfw/

[144] Taigman, Y., and Wolf, L., "Leveraging Billions of Faces to Overcome Performance Barriers in Unconstrained Face Recognition," http://arxiv.org/pdf/1108.1122.pdf

[145] Ibid.

[146] Bohannon, J., "Facebook Will Soon Be Able To ID You In Any Photo", *Science*, February 5, 2015. http://news.sciencemag.org/social-sciences/2015/02/facebook-will-soon-be-able-id-you-any-photo

to monitor the outgoing calls of suspects. And the USA PATRIOT Act grants explicit permission to the government to intercept wire, oral, and electronic communications if they relate to terrorism. It too would seem logical that one has no legitimate belief that retail records are kept private. ... If there is no expectation that one's purchases are being kept private, then the government can scrutinize these records without a warrant. Likewise, under the USA PATRIOT Act, the government is granted permission to search such records. Thus, because there is no expectation that retail records are private, and because of the USA PATRIOT Act's authority, the government's use of data mining with retail records does not seem to currently violate any federal laws on privacy."[147]

Ultimately, data mining that includes face recognition from digital still pictures and videos might be used to show unusual patterns of behavior and thus provide one more input to a new anti-crime algorithm: inferences of intent of specific individuals.

The message is not just that there will be a lot of data and video on file about everyone--we know that already--it is that more about you will be known than is contained in the databases. The synergy of the data, the relationships of one source to another, the correlations among the data and their trends tell more about you than any single source. The whole picture will be greater than any of its parts. It could produce estimates of likelihood of future actions.

The synthesis of these data will be difficult and probably beyond human capacity. Can it be automated? The IBM computer Watson is now being used in the analysis of large-scale databases in fields of finance and health. Perhaps it (or its progeny) can be reprogrammed from Watson to Sherlock and its inferential power harnessed to identify the risk associated with specific people.[148] Trusting this system could be very dangerous because Sherlock may be Big Brother in disguise. Sherlock offers a Faustian compromise of privacy and it must be near perfect since false positives will bring lasting and erroneous stigma, while false negatives will mean potential LWs and SIMADs could remain undetected.

5.3 Monitoring purchases of critical materials

A logical strategy in identifying potential lone wolf terrorists is to monitor the purchasing of critical materials. What materials? Anything that might be used as a weapon or in constructing weapons including, for example, large quantities of chemicals useful in making explosives or incendiary devices, biological materials and laboratory instruments, and poisons or components of poisons.

---

[147] Persky, D., "Common Materials Turned Deadly: How Much Does America Have to Monitor to Prevent Further Acts of Terrorism?" *National Security Law Brief* 4, no. 1 (2013): 43-56.
[148] IBM Watson http://www-03.ibm.com/press/us/en/presskit/27297.wss

Timothy McVeigh, the convicted and executed lone wolf bomber of the federal building in Oklahoma City in 1995, constructed a large explosive device using 5,000 pounds of ammonium nitrate, an agricultural fertilizer, exploding targets, and rocket fuel and nitromethane, a racing car fuel, in the back of a rented pickup truck used for the delivery of the bomb. One hundred sixty eight people, including 19 children were killed in the blast.[149]

The Norwegian lone wolf, Anders Breivik, is thought to have used the same materials as McVeigh.[150]

Jose Pimentel, the New York Lone Wolf who was apprehended before he could act said that he bought his bomb-making materials at Home Depot and other stores, varying the point of purchase to avoid suspicion.[151]

Jordan Gonzalez pleaded guilty to attempting to produce ricin and abrin for weapons using the seeds from plants from which these toxins could be derived. According to the FBI he bought thousands of seeds to produce the toxins and had "filtering equipment, respirators, glass vials, a spraying device, and projectile ... Gonzalez made the purchases through an online third-party vendor marketplace, and all the items were delivered to him at his Jersey City apartment."[152]

The problem becomes much more difficult when the quantities required are small but more easily obtained. For example, the Boston Marathon Bombers used pressure cookers to contain their improvised explosive devices (IED); the U.S. Department of Homeland Security had warned about conversion of pressure cookers into IEDs. They said:

"Typically, these bombs are made by placing TNT or other explosives in a pressure cooker and attaching a blasting cap at the top of the pressure cooker. The size of the blast depends on the size of the pressure cooker and the amount of explosive placed inside. Pressure cooker bombs are made with readily available materials and can be as simple or as complex as the builder decides. These types of devices can be initiated using simple electronic components including, but not limited to, digital watches, garage door openers,

---

[149] Eddy, M., Lane, G., Pankratz, H., and Wilmsen, S., "Guilty on Every Count", *Denver Post Online* June 3, 1997 McVeigh was found guilty of murder and executed in 2001. The charges against McVeigh included "conspiracy to use a weapon of mass destruction, using a weapon of mass destruction and destroying a federal building".

[150] "Oslo Norway Bombing: Suspect Anders Behring Brevik Bought Tons Of Fertilizer, Wrote Manifesto", *The Huffington Post*, http://www.huffingtonpost.com/2011/07/23/oslo-bombing-anders-behring-brelvik-norwegian-suspect-fertilizer_n_907697.html

[151] CBS: terror suspect allegedly plotted bombings against US troops, NYC sites. http://newyork.cbslocal.com/2011/11/20/bloomberg-lone-wolf-arrested-on-terror-charges/

[152] "New Jersey Pharmacist Admits Attempting to Weaponize Deadly Toxins and Possessing Narcotics Manufacturing Equipment", US Attorney's Office, May 29, 2014. http://www.fbi.gov/newyork/press-releases/2014/new-jersey-pharmacist-admits-attempting-to-weaponize-deadly-toxins-and-possessing-narcotics-manufacturing-equipment

cell phones or pagers. As a common cooking utensil, the pressure cooker is often overlooked when searching vehicles, residences or merchandise crossing the U.S. Borders... (this is) a technique commonly taught in Afghan terrorist training camps."[153]

In 2011, the U.S. Department of Homeland Security announced plans to regulate the sale of ammonium nitrate. Sale of more than 25 pounds requires registration with the Department and records of sales must be maintained for at least 2 years.[154]

The Office of Homeland Security in the United States has an active bomb making materials awareness program (BMAP), developed in conjunction with the Federal Bureau of Investigation. The purpose of the program is to:

"... help employees more easily identify homemade explosives precursor chemicals and improvised explosive device (IED) components, and recognize suspicious purchasing behavior that could indicate potential bomb-making activities.... Powerful explosives can be made from precursor chemicals found in common consumer goods that are readily available commercially, making them highly attractive to terrorists attempting to avoid the obstacles to obtaining conventional explosives. Homemade explosives were used in several high-profile incidents, such as the 2005 London transit attacks, the 2001 Richard Reid "shoe bomb" plot, and the 1995 Oklahoma City bombing. [...]

The importance of educating employees at the point of sale cannot be overstated, as this is the best way to ensure early detection of the sale of precursor chemicals to suspect individuals. Employees' ability to recognize and report behaviors indicative of potential homemade explosive-related activity is crucial to the possible prevention of an IED attack."[155]

Finding buyers of critical materials by lone wolves becomes very difficult when the amount of material involved is small (exotic toxins or biological components, for example), when the materials have multiple uses (dynamite), when the terrorist is an employee of the manufacture of the material (as is suspected in the anthrax killings), or as is often the case, when the terrorist uses Tor--the hidden Internet--for their transactions.

Tor is an acronym for "the onion project," a hidden version of the Internet that permits almost total anonymity to users. It is called "the onion project" because the communication occurs in layers and the hopping from one server to another within each layer is randomly determined. Surprisingly, it was initiated in 1995 by the U.S. Naval Research Laboratory and was expanded

---

[153] Information Bulletin, US DHS, "Potential Terrorist Use Of Pressure Cookers."
[154] Homeland Security Plans To Regulate Bomb Fertilizer, Los Angeles times, August 2, 2011. http://articles.latimes.com/2011/aug/02/nation/la-na-ammonium-nitrate-20110803
[155] Bomb-Making Materials Awareness Program, Department of Homeland Security http://www.dhs.gov/bomb-making-materials-awareness-program

in 2001 by two MIT students. Since 2006 it has operated as a nonprofit corporation and is supported today with funding from the Navy, the U.S. National Science Foundation, the U.S. Department of State and the Broadcasting Board Of Governors.[156]

> "Tor passes your traffic through at least 3 different servers before sending it on to the destination. Because there's a separate layer of encryption for each of the three relays, Tor does not modify, or even know, what you are sending into it. It merely relays your traffic, completely encrypted through the Tor network and has it pop out somewhere else in the world, completely intact. The Tor client is required because we assume you trust your local computer. The Tor client manages the encryption and the path chosen through the network. The relays located all over the world merely pass encrypted packets between themselves."[157]

Tor is free and legal everywhere in the world. Legitimate uses for Tor include communications by people and businesses that want to keep business strategies confidential, by activists and whistleblowers to report on abuses, by journalists protecting their sources and by military and law enforcement agencies to protect intelligence gathering activities. But it is the shady uses for Tor that concern us here.

Tor has carried child pornography (and has been attacked by "hactivists"), and has had drug dealing and gambling sites. It is also well-known for its underworld marketplaces. One of the most notorious was Silk Road.

> "A Boston Globe editor and reporter were able to access the site after downloading Tor Project software and registering as Silk Road users. Described as an 'anonymous marketplace,' the site promotes its ability to 'protect your identity through every step of the process, from connecting to this site, to purchasing your items, to finally receiving them,' through the use of Tor technology. The homepage features pictures of various drugs for sale--including heroin and cocaine--and allows buyers to place them in a shopping cart, similar to those on Amazon and other consumer sites. In addition to drugs, the site purports to provide access to other illegal products, including forged documents, and links to a separate marketplace called the Armory, designated for 'small arms weaponry for the purpose of self defense' "[158]

---

[156] McKim, Jenifer B. (8 March 2012). "Privacy software, criminal use". The Boston Globe. http://wayback.archive.org/web/20120312225054/http://articles.boston.com/2012-03-08/business/31136655_1_law-enforcement-free-speech-technology/2

[157] Tor, Anonymity on Line: https://www.torproject.org/docs/faq.html.en

[158] McKim, J., "Walpole Company's Anonymity Software Aids Illicit Deals," *The Boston Globe*, March 8, 2012. http://www.bostonglobe.com/business/2012/03/08/walpole-company-anonymity-software-aids-elicit-deals/n1icZ1d30WjvUmmQqS7vjM/story.html

Law enforcement authorities closed down Silk Road in October 2013.[159] It is reported to have had 900,000 registered users[160]; its sales were estimated to be 9.5 million Bitcoins and its revenue, 600,000 Bitcoins. This translates to $1.2 billion in sales and $80 million in income.[161] Silk Road 2.0 was soon operating after the original Silk Road was closed down, but it too was soon closed. The following image shows what appears on a computer screen when a closed-down site is retrieved:

U.S. Immigration and
Customs Enforcement

THIS HIDDEN SITE HAS BEEN SEIZED

as part of a joint law enforcement operation by
the Federal Bureau of Investigation, ICE Homeland Security Investigations,
and European law enforcement agencies acting through Europol and Eurojust

in accordance with the law of European Union member states
and a protective order obtained by the United States Attorney's Office for the Southern District of New York
in coordination with the U.S. Department of Justice's Computer Crime & Intellectual Property Section
issued pursuant to 18 U.S.C. § 983(j) by the
United States District Court for the Southern District of New York

One can also buy biological laboratories and biological parts (e.g. promoters, proteins, DNA, RNA, coding sequences, plasmids, terminators and other genetic molecular components) from legitimate open sources on the Internet. [162]

In May 2014, the FBI arrested Ryan Chamberlin in San Francisco. According to a local report, when they searched his apartment, apparently based on a tip, they found bomb-making material and "biological agents and lethal toxins" that he had allegedly purchased on the darknet from a site called Black Market Reloaded (which has reportedly been shut down). [163] [164]

---

[159] Flitter, E., "FBI Shuts Alleged Online Drug Marketplace, Silk Road," *Yahoo News*, August 2013; http://news.yahoo.com/fbi-raids-alleged-online-drug-market-silk-road-153729457.html
[160] Ibid.
[161] Hill, K., "The FBI's Plan for the Millions Worth of Bitcoins Seized From Silk Road," *Forbes*, October 4, 2013. http://www.forbes.com/sites/kashmirhill/2013/10/04/fbi-silk-road-bitcoin-seizure/
[162] See: Registry of Standard Biological Parts: http://parts.igem.org/Catalog
[163] Tiku, N., "FBI alleges San Francisco PR consultant bought toxins on the darknet," ValleyWag, June 2, 2014. http://valleywag.gawker.com/fbi-alleges-san-francisco-pr-consultant-bought-toxins-o-1589315960
[164] FBI, U.S. Attorney's Office, "San Francisco Resident Indicted for Possession of an Improvised Explosive Device" June 13, 2014: http://www.fbi.gov/sanfrancisco/press-releases/2014/san-francisco-resident-indicted-for-possession-of-an-improvised-explosive-device

Could a darknet site be set up as a honeypot [165] that is a place that draws potential lone wolves or SIMADs, who are trying to obtain materials, funding, information, or alliances? One source claims:

> "Another approach officials have taken is to create phony terrorist websites. These can spread disinformation, such as instructions for building a bomb that will explode prematurely and kill its maker or false intelligence about the location of U.S. forces in Iraq, intended to lead terrorist fighters into a trap." [166]

In an age of SIMAD, it may become imperative use such strategies and to track transactions on sites where buyers and sellers believe they have anonymity.

5.4 Monitoring "friends" and communications on social media and email

Some thieves who were not very smart have posted their photos on Facebook with their booty, one even included his new address, employer, and working hours. One bank robber, a 19 year old blond woman is supposed to have displayed the stack of cash she stole on YouTube (and was arrested shortly thereafter). Steven Kazmierczak, the Northern Illinois University shooter who killed 5 people and himself and wounded 21 had a black T-shirt that said, "TERRORIST in white letters and a red graphic of an AK-47 assault rifle." He wore it to the shooting. [167] [168] News reports allege that Christopher Cornell wrote material on Twitter that showed his support for the Islamic State; he apparently developed a plot to bomb the Capitol and shoot the persons fleeing from the blast, with an FBI informant who was posing as an accomplice. [169] [170]

Potential lone wolves can leave a different kind of footprint. Some potential LWs display their frustrations and even their plans on social media; anticipating their crimes requires only reading their posts. Jose Pimentel, for example, arrested on charges of planning to bomb targets in New York City, managed his own radical website, TrueIslam1.com. This website "hosted an impressive archive of jihadist texts, with audio and video organized by means of the online

---

[165] A honey pot is similar to a sting operation in which an attractive object is displayed for the purpose of finding out who is interested in that object.

[166] Kaplan, E., "Q&A: Terrorists on the Internet", *New York Times*, March 6, 2006.
http://www.nytimes.com/cfr/international/slot2_030606.html?_r=0

[167] "Portrait of the School Shooter as a Young Man", Esquire.com, February 12, 2009.
http://www.esquire.com/features/steven-kazmierczak-0808

[168] Murderpedia: http://murderpedia.org/male.K/k/kazmierczak.htm

[169] Knickerbocher, B., "How Alleged Lone Wolf Terrorist Plotted Attack On US Capitol – And Was Stopped", *Christian Science Monitor*, January 15, 2015 http://www.csmonitor.com/USA/Justice/2015/0115/How-alleged-lone-wolf-terrorist-plotted-attack-on-US-Capitol-and-was-stopped

[170] RT: "Capitol Hill Terror Suspect Says He Would Have Shot Obama": March 7 2015: http://rt.com/usa/238609-capitol-terror-suspect-obama/

---

publishing tool Blogger. The website connects to Pimentel's YouTube channel, which was similarly thorough; it had collected more than 600 videos relating to radical and violent interpretations of Islam, 60 of which he had uploaded himself. This channel had more than 1,500 subscribers."[171]

Anders Breivik published a 1,500 page manifesto on the day of his killing spree in Norway in 2011. The Holocaust Museum shooter, James von Brunn ran an anti-Semitic web site called the Holy Western Empire. Major Nidal Malik Hasan developed a Power Point presentation that was also online, called "The Koranic World View as it Relates to Muslims in the U.S. Military." [172]

Sefa Riano in Indonesia planned to bomb the Myanmar Embassy but his plans appeared on his Facebook page and this apparently led to his apprehension.[173]

Information about prospective lone wolves might be gained not only by reading what has been uploaded, but also by observing who has visited or downloaded lone wolf relevant sites. The article "How to Make a Bomb in Your Mom's Kitchen,"

"...was downloaded by individuals who plotted terrorist attacks in both the U.S. and the UK (e.g., Naser Jason Abdo, a Muslim U.S. soldier who allegedly plotted to attack [a restaurant near] the Fort Hood military base, and [Jordan] Pimentel, who had started making a pipe bomb based on the recipe when he was arrested). The magazine [Inspire] has also heavily featured the writings of a Syrian al-Qaeda-linked terrorist, Abu Musab al-Suri, who pioneered the concept of individual terrorism before the 9/11 attacks. ...The articles in Inspire clearly promote individual jihad; thus, the Fall 2010 edition editorialized: Spontaneous operations performed by individuals and cells here and there over the whole world, without connections between them, have put the local and international intelligence apparatus in a state of confusion, as arresting the members of aborted cells does not influence the operational activities of others who are not connected with them." [174] [175]

The ideas and methods for terror attacks are meant for anyone, including those without direct ties to al-Qaeda or its affiliates. Thus, the Summer 2010 issue advised making a pipe bomb using

[171] Weimann, G., "Lone Wolves in Cyberspace", *Journal of Terrorism Research*, Vol3, Issue 22 (2012) http://ojs.st-andrews.ac.uk/index.php/jtr/article/view/405/431
[172] Ibid.
[173] Karmini, N., Riano, S., "Indonesian Terror Suspect, Caught Through Facebook Posts", *Huffingtonpost*, August 20, 2013 http://www.huffingtonpost.com/2013/06/20/sefa-riano-caught-through-facebook-indonesia_n_3471594.html
[174] Op cit., Weimann, G.
[175] FBI: Naser Jason Abdo Sentenced to Life in Federal Prison in Connection with Killeen Bomb Plot: http://www.fbi.gov/sanantonio/press-releases/2012/naser-jason-abdo-sentenced-to-life-in-federal-prison-in-connection-with-killeen-bomb-plot

everyday materials, the Fall 2010 issue encouraged using one's car to "mow down" people in crowded places and the Winter 2010 issue discussed how to blow up buildings.[176]

A new software tool has been developed for law enforcement that searches for communications that hint at gang violence, drug dealing, crimes against children and human trafficking. The tool, called *Social Media Monitor,* was developed by LexisNexis. It can:

> "...alert officers to potential areas of concern and help them identify posts or tweets within specific geographic locations. By entering a few search terms, law enforcement personnel are provided with a social canvas within minutes, adding a virtual dimension to traditional public records data. In a recent demonstration, LexisNexis officials showed how monitoring the Twitter feeds of gang members could help them learn code words, drug drops, meeting locations and criminal trends within cities or even larger demographic areas."[177]

"Geo-fencing" has already proved to be effective in investigations. Toronto Deputy Chief Peter Sloly described how his department used geo-fencing[178], a tool "to search for open-source social media content within defined geographical areas and time frames." Geo-fence can be created anywhere and for any extent. He gives the example of a stabbing occurred during a large summer festival. The geo-fence and scanning of social media posts posted within minutes of the incident provided "more information [...] than we would have from having 30 officers canvass the area of the scene for three days. We were also able to keep an eye on social media for possible revenge crimes and know the hot spots where those might occur."[179]

A paper presented at the European Intelligence and Security Informatics Conference in 2012 outlined some methods for analyzing Internet content to collect weak signals of terrorist intent. Their "techniques for harvesting and analyzing data from the Internet" to collect weak signals is based on an automated analysis of content of contributions to interesting web sites; the analysis is designed to bring key suspects to the surface for human examination. The sites used in the automated analysis are identified through tracing hyperlinks connecting the sites flagged through the use of keywords and including searches of the Internet not accessed by existing search engines (e.g. Google). On these sites, natural language processing is used to identify which sites, contributions, and contributors are worth taking a deeper look at. They break the identification

---

[176] Ibid

[177] Police Get Tool For Patrolling Social Media (Nov, 2013): http://gcn.com/blogs/emerging-tech/2013/11/social-media-monitor.aspx

[178] "Use of Social Media For Investigation and Crime Prevention", a chapter in *The Role of Local Law Enforcement Agencies in Preventing and Investigating Cybercrime*, pp 39, April 2014.
http://www.policeforum.org/assets/docs/Critical_Issues_Series_2/the%20role%20of%20local%20law%20enforcement%20agencies%20in%20preventing%20and%20investigating%20cybercrime%202014.pdf
[179] Ibid.

problem down into a number of smaller sub-problems "such as identifying motives (intent), capabilities, and opportunities."[180]

Who is "friends" with whom? Do criminals tend to hang out with other people like themselves on social networks? If so, then keeping track of networks of like people could provide another source of information about proto lone wolves. Similarly, early detection strategies should include simply reading social media sites to find threatening uploads, and checking who has downloaded this material. It seems counterproductive but the list of key words used by analysts in Homeland Security to search social media for messages containing hints of terrorism-to-come has been published. The Electronic Privacy Information Center--a group that focuses on First Amendment rights and privacy issues filed a Freedom of Information Act request and obtained the release of the list.[181] It contains over 300 words such as "national laboratory", "infection", "agro terror", "subway", "cyber command", "bridge", and "nerve agent".

5.5 Third-party reporting of unusual behavior

Many states in the U.S. require mandatory reporting of certain kinds of behavior even if this reporting requires breech of usual rules of confidentiality. In California, for example, if a teacher, coach, social worker, policeman or fireman, family or drug councilors, or computer technician has a reasonable suspicion that a child has been abused or neglected, they are required by law to submit a report. In Massachusetts, doctors, dentists, nurses, priests, rabbis and mental health professionals are included. Similar laws exist in many states to protect elderly persons from abuse.[182] In addition, in the United States, the Bank Secrecy Act requires financial institutions (e.g. banks, brokerage houses, currency exchange bureaus, insurance companies, etc.) to file "suspicious activity reports" with the Financial Crimes Enforcement Network, an agency of the U.S. Department of Treasury.[183] The USA Patriot Act of 2001 requires banks to report large

---

[180] Brynielsson, J., Horndahl, A., Johansson, F., Kaati, L., Martenson, C., Svenson, P., "Analysis Of Weak Signals For Detecting Lonewolf Terrorists," Swedish Defense Research Agency, 2012 European Intelligence and Security Informatics Conference, 2012
http://www.foi.se/global/our_knowledge/decision_support_system_and_information_fusion/foi-s--4093--se.pdf
[181] Miller, D., "Revealed: Hundreds of words to avoid using online if you don't want the government spying on you (and they include 'pork', 'cloud' and 'Mexico'), Daily Mail, 26 May, 2012; http://www.dailymail.co.uk/news/article-2150281/REVEALED-Hundreds-words-avoid-using-online-dont-want-government-spying-you.html#ixzz1w1SkH6gY
[182] The California Child Abuse And Neglect Reporting Law:
http://www.dominican.edu/academics/resources/facultyresources/file/childabuselaw.pdf Also: Miller, M., "Mandated Reporting of Suspected Child Abuse", *Board of Psychology Update*, January 2008. http://www.girlsinc-alameda.org/files/MandatedReportingSuspectedChildAbuse.pdf
[183] Electronic Filing Instructions, Financial crimes enforcement network, October 2012:
http://www.fincen.gov/forms/files/FinCEN%20SAR%20ElectronicFilingInstructions-%20Stand%20Alone%20doc.pdf

---

financial transactions to help detect money laundering; it also requires notification of transfers of biological agents and toxins that have terrorist potential. Many professionals (e.g. psychiatrists, physicians, psychologists, lawyers, counselors, etc.) in many jurisdictions are required--both by law and professional ethical standards--to report certain potentially threatening behavior to authorities.

In the United States, there are several programs that encourage third party reporting of unusual behavior. There is a public awareness program called "If you see something, say something." There are well-established neighborhood watch programs in which residents are encouraged to become the eyes and ears of law enforcement.

Department of Justice, the Department of Homeland Security and the Federal Bureau of Investigation cooperate in a Nationwide Suspicious Activity Reporting Initiative.[184] The incident reporting system has special instructions for chemical security, cyber security, emergencies in federal buildings, oil and chemical spills, criminal activity, and terrorist activity. An online form (not anonymous) is provided to report on suspected terrorist activities.

In 2012, a sample of over 1,500 adults in the United States was polled to determine their attitudes about reporting certain kinds of behavior to the police. Participants said they would be "very likely" or "somewhat likely" to call the police if they heard or saw a person:

- Talking about planting explosives   89.2%
- Talking about breaking into a house   88.5%
- Traveling overseas to join a terrorist group   75.4%
- Distributing handouts in support of terrorism 74.6%
- Talking about joining a terrorist group   70.1%
- Reading material from a terrorist group   49.1% [185]

There are several examples of LW terrorists being captured as a result of third party information. Ted Kaczynski was captured when his brother recognized his writings in his manifesto.[186] Faisal Shahzad's plot to explode a bomb in an SUV in Times Square was foiled when a street vendor saw something and said something; he saw smoke in the vehicle and alerted police, who immediately evacuated the area. The bomb never ignited and no one was hurt."[187] According to

---

[184] Report Suspicious Activity. Homeland Security: http://www.dhs.gov/how-do-i/report-suspicious-activity
[185] Testimony of William Braniff, Executive Director National Consortium for the Study of Terrorism and Responses to Terrorism (START), to the House Committee on Homeland Security Subcommittee on Oversight and Management Efficiency "Why Can't DHS Better Communicate with the American People?"
[186] Plinkington, E., "My Brother the Unabomber," *The Guardian,* Sept 15, 2009.
http://www.theguardian.com/world/2009/sep/15/my-brother-the-unabomber
[187] Neumeister, L., "Times Square Bomber Indicted with 10 Terrorism, Weapons Charges", *Huffington Post,* June 17, 2010; "http://www.huffingtonpost.com/2010/06/17/times-square-bomber-indic_n_616534.html

the FBI, Khalid Ali-M Aldawsari was preparing to make bombs. and a suspicious supplier of one of the chemicals that Aldawsari ordered called the police.[188] In the case of Sami Osmakac, "A confidential source allegedly told federal officials in September 2011--the 10th anniversary of 9/11--that Osmakac wanted al-Qaeda flags."[189]

The darkside of third party reporting is clearly illustrated in its use by Nazis in the Second World War to find Jews; they required citizens in Germany and in occupied countries to report Jews that had not been otherwise identified. Some few countries to their great credit, did not comply with demands of the Gestapo: among these Denmark stands out.[190] And certainly there were courageous individuals who helped Jews escape the fate that the Nazis had prepared for them, but there is no doubt that most countries and people under German control joined in the Jew-hunt. Several books document this behavior in, for example, France and Poland. [191]

In the U.S., in the 1950s, Senator Joseph McCarthy through demagoguery and Cold War paranoia was able to induce many Americans to report their colleagues and associates as "sympathizers" who were soft on Communism and "fellow travelers." It was not the country's proudest hour. Earlier witch-hunts also come to mind where people pointed their fingers at neighbors based on ill-informed hysteria, paranoia, and fear. If third party reporting is to play a role in identifying possible lone wolves or SIMADs, as seems likely, let us also introduce checks and balances that preserve justice.

5.6 Screening

A lot of screening goes on today; sometimes the screens are initiated by the persons who will be screened and other times by third parties. Some screens are based on person to person interviews (such as interviews by psychiatrists who are screening for depression) and others are based on observable physical characteristics including gait, odor, and facial expression which can be "read" to detect mood.[192] Probably the most familiar screen is at airports where imaging

---

[188] "Texas Resident Arrested on Charge of Attempted Use of Weapon of Mass Destruction," FBI, http://www.fbi.gov/dallas/press-releases/2011/dl022411.htm

[189] "'Once I have this, you can take me in five million pieces': Kosovan immigrant, 25, to Florida 'planned to cause havoc with car bomb, machine guns and explosives belt'" Daily Mail, January 9, 2012. http://www.dailymail.co.uk/news/article-2084327/Sami-Osmakac-planned-cause-havoc-Florida-car-bomb-assault-rifle-explosives-belt.html?ito=feeds-newsxml

[190] Ignatieff, M., "One Country Saved Its Jews. Were They Just Better People? The Surprising Truth About Denmark In The Holocaust", New Republic, December 2013. http://www.newrepublic.com/article/115670/denmark-holocaust-bo-lidegaards-countrymen-reviewed

[191] Tzur, N., "New Book Examines Poles Who Killed Jew's During WWII", The Times Of Israel, August 28 2014

[192] Scharr, J., "Facial-Recognition Tech Can Read Your Emotions, livescience, "January, 2014; http://www.livescience.com/42975-facial-recognition-tech-reads-emotions.html

machines and physical searches are used to find prohibited weapons and materials. Some airports use profiling as well. Prospective employers sometimes hire firms to conduct background checks of applicants using social media screens. There are screens conducted for medical reasons too: such as checks for SARS, Ebola, and Avian flu, and patient initiated screens for genetic or other indications of possible diseases or propensity to fall ill to diseases in the future such as cancer, ALS, diabetes, and high blood pressure. Amniocentesis is used to detect the health and gender of fetuses. Mental health screening is available also for signs of PTSD, potential for suicide, anxiety, depression, and alcoholism. But for our purposes here, we ask: "Can effective psychological or physical screens be created to detect lone wolves and SIMADS before they act?"

The airport screening systems have been fairly successful. Air travelers are not only screened at airports for concealed weapons, but for outward appearances that seem to be discordant, or for providing inconsistent answers to questions posed by police or other officials. Screening at Ben Gurion Airport near Tel Aviv, Israel, includes, in addition to the now standard searches of baggage and machine searches of persons, ethnic profiling, which subjects people with Arab appearance or names to more rigid scrutiny. That activity has aroused some civil rights complaints. But it is defended by Ariel Merari, an Israeli terrorism expert who said that ethnic profiling is both effective and unavoidable: "It's foolishness not to use profiles when you know that most terrorists come from certain ethnic groups and certain age groups," he said. "A bomber on a plane is likely to be Muslim and young, not an elderly Holocaust survivor. We're talking about preventing a lot of casualties, and that justifies inconveniencing a certain ethnic group."[193]

There are research efforts underway to develop and implement real-time screening systems of this sort. For example, in the United States, the Homeland Security Advanced Research Agency and The Science And Technology Human Factors Behavior Science Division Of The Department of Homeland Security are developing the FAST system that is intended to search for individuals with malintent, as though malintent were a disease and the factors involved in the screen were symptoms of that disease:

> "… the specific focus of FAST is identifying individuals who exhibit physiological indications, which in the specific screening settings, are determined to be associated with malintent. Behavioral scientists hypothesize that someone with malintent may act strangely, show mannerisms out of the norm, or experience extreme physiological reactions based on the extent, time, and consequences of the event. The FAST technology design capitalizes on these indicators to identify individuals exhibiting characteristics associated with malintent." [194]

[193] "Rights Group Challenges Israel's Airport Security", *NBC news*, March 19, 2008.
http://www.nbcnews.com/id/23714853/ns/world_news-mideast_n_africa/t/rights-group-challenges-israels-airport-security/
[194] Future Attribute Screening Technology Project, US Department Of Homeland Security, December, 2008:
https://www.dhs.gov/xlibrary/assets/privacy/privacy_pia_st_fast.pdf

The malintent indicators with which FAST is experimenting include:

1. A remote cardiovascular sensor (heart rate, heart rate variability, respiration rate, and arrhythmia).
2. A remote eye tracker (the eye position and pupil diameter).
3. Cameras for thermal properties of skin and eye movements.
4. High resolution video for facial features
5. An audio system for human voice for pitch change.
6. Other sensor types such as for pheromones detection

In 2008, when the project was in an early stage, a Homeland Security spokesperson said "We are running at about 78% accuracy on malintent detection, and 80% on deception."[195] While the design and findings of this experiment were challenged, they nevertheless are remarkable and raise the questions associated with false positives and negatives. As noted in Section 5.2, false positives mean that innocent people have been identified as potential terrorists and will have to prove their innocence and suffer unjustified stigma, and false negatives mean that some potential lone wolves have beaten the system.

Psychological screening tools also give some hope of identifying potential psychopaths (and certainly some LWs are psychopaths) through semi-structured interviews. One such tool, the Hare Psychopathy checklist, assesses factors correlated to interpersonal and affective deficits as well as symptoms of antisocial behavior. Some of the factors assessed include:

- Glibness
- Grandiose self-worth
- Pathological lying
- Lack of remorse
- Lack of empathy
- Irresponsibility
- Delinquency
- Promiscuity
- Need for stimulation
- Parasitic lifestyle
- Impulsivity

Professor Kent Kiehl, of the University of New Mexico, applies the Hare Psychopath Checklist in his studies. He says:

"The (Hare test) scores range from zero to 40. The average person in the community, a male, will score about 4 or 5. Your average inmate will score about 22. An individual

---

[195] Future Attribute Screening Technologies Precrime Detector, technovelgy.com: http://www.technovelgy.com/ct/Science-Fiction-News.asp?NewsNum=1891

with psychopathy is typically described as 30 or above. Brian (Dugan, a man serving 2 life sentences for raping and murdering a 10 year old girl) scored 38.5 basically. He was in the 99th percentile." [196]

So, apparently, there is some ability for tests such as the Hare Psychopath Checklist to discriminate the possibility of abnormal violent behavior and potential danger, at least after the fact. But there is disagreement about the validity of such tests and their ability to forecast future behavior.

And keep in mind that some LWs appear normal, placid, friendly, happy, and well adjusted. A girl friend of Steve Kazmierczak, the Northern Illinois University shooter, said, "He was probably the nicest, most caring person ever... He was just under a lot of stress from school. He didn't have a job, so he felt bad about that -- he wasn't erratic, he wasn't psychotic, he wasn't delusional, he was Steve. He was normal."[197]

5.7 Brain physiology

There has been great progress in this field since the turn of this century, primarily using brain imaging approaches and electroencephalography. Imaging is accomplished with functional magnetic resonance imaging (fMRI) and positron emission tomography (PET). Electroencephalography (EEG) uses brain generated electrical currents on the surface of the scalp to detect and measure brain activity. The imaging techniques depend on changes in blood flow in the brain: when a portion of the brain is active it consumes more oxygen and blood flow increases. Imaging techniques can detect this activity and identify the active anatomical regions in which certain thought processes occur. Neuroscientists using this brain imaging have aided in the selection of jurors and in the analysis of discussions in focus groups. EEG is often used in the diagnosis of brain disorders. MRI and PET scanning systems involve big machines; EEG can be miniaturized and has been used in game interfaces in consumer applications costing less than $100.[198]

What if, under benign interrogation, an interviewer, or even a computer program, were to toss out certain charged words, or present a graphic terrorist scenario and judge the response using brain images or EEG responses? Could this serve as the basis for an advanced lie or malintent detector? One scientific paper, reporting on an fMRI experiment says:

---

[196] Bradley Hagerty, B., *Inside A Psychopath's Brain: The Sentencing Debate*, June 30, 201012.
http://www.npr.org/templates/story/story.php?storyId=128116806
[197] *Portrait of the School Shooter as a Young Man*. Esquire.com, February 12, 2009.
http://www.esquire.com/features/steven-kazmierczak-0808
[198] "Brain Control Technology - A New Olympic Sport?", *Enhanced Online News*, February 23, 2010.
http://www.enhancedonlinenews.com/portal/site/eon/permalink/?ndmViewId=news_view&newsId=20100223007492&newsLang=en

---

"Lie was discriminated from truth on a single-event level with an accuracy of 78% [...] our findings confirm that fMRI, in conjunction with a carefully controlled query procedure, could be used to detect deception in individual subjects."[199]

Another researcher reported, "Differences in EEG patterns among psychiatric patients, prison inmates, and normal controls have been demonstrated by use of electronic frequency analysis of electroencephalograms. Some correlations between personality characteristics, as reflected in the (correlation between) Rorschach test, and EEG frequency patterns are reported."[200]

Figure 5.1 shows two brain fMRIs[201]. The one on the right, of a person deemed to be normal; the image on the left is a brain scan of a convicted murderer who exhibited violent and bizarre behavior after suffering a head injury at age 20.[202] Impulsive behavior is apparently controlled by the brain's prefrontal cortex and the person whose brain scan is shown on the right in the figure below was thought to have been injured in this area.

Figure 5.1 Comparative Image of two brain fMRIs

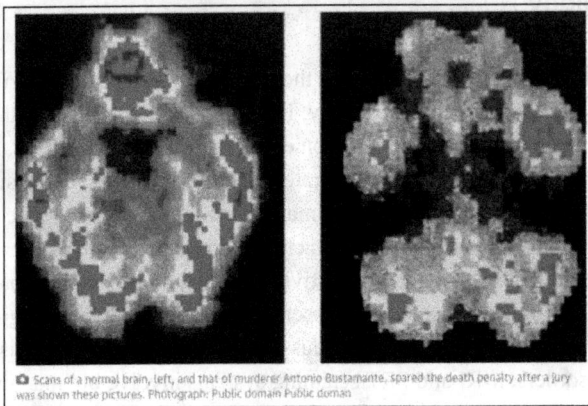

Scans of a normal brain, left, and that of murderer Antonio Bustamante, spared the death penalty after a jury was shown these pictures. Photograph: Public domain Public domain

Source: *How to spot a murderer's brain* http://www.theguardian.com/science/2013/may/12/how-to-spot-a-murderers-brain

---

[199] Langleben DD, Loughead JW, Bilker WB, Ruparel K, Childress AR, Busch SI, Gur RC., *Telling Truth From Lie In Individual Subjects With Fast Event-Related fMRI*. Hum Brain Mapp. 2005 Dec;26(4):262-72.
http://www.ncbi.nlm.nih.gov/pubmed/?term=Telling+truth+from+lie+in+individual+subjects+with+fast+event-related+fMRI
[200] Rabinovitch, M. Sam; Kennard, Margaret A.; Fister, W. P. "Personality correlates of electroencephalographic patterns: Rorschach findings." *Canadian Journal of Psychology Revue Canadienne de Psychologie*, 1955, 29-41.
https://www.researchgate.net/publication/9232327_Personality_correlates_of_electroencephalographic_patterns_Rorschach_findings
[201] Adams, T., "How to Spot a Murderer's Brain," *The Guardian*, May 11, 2013:
http://www.theguardian.com/science/2013/may/12/how-to-spot-a-murderers-brain
[202] "The Anatomist of Crime", *The Pennsylvania Gazette*, October 28, 2013. http://thepenngazette.com/the-anatomist-of-crime/

But as critics have pointed out there is a long way to go. The relationship between brain structure, activity, and thought may vary with several factors such as handedness and experience, and increasing oxygen uptake prior to or during the test may fool the whole process. [203]

5.8 Genetics

Why are some people violent? Is there such a thing as a gene that promotes aggression? Can propensity to violence be an inherited trait? Attempts to answer such questions tangle with politics as well as science. In 1993, Dr. Louis W. Sullivan, who was then the secretary of Health and Human Services in the Bush administration, proposed a plan to allocate $400 million dollars to the study of youth violence. Many critics of the proposal claimed it was biased and that it was an attempt "to prove that minorities are biologically prone to commit violence. Others said treating violence as a health issue obscured the roots of urban upheaval--drugs, poverty, bad schools and the breakdown of the family."[204] The arguments were so intense that the General Accounting Office and a special commission looked into the possibility of bias; each found no such intent. Yet in 1993, the issue was (and still is) boiling and the Clinton Administration dropped the proposal.[205] Some people saw the proposed research as being racist and the arguments about nature (genetics) vs. nurture (childhood and parental influence, poverty, and education) persist. But there is increasing clarity.

In 2002, in a groundbreaking study, Dr. Terrie Moffitt and her colleagues:

"… studied a large sample of male children from birth to adulthood to determine why some children who are maltreated grow up to develop antisocial behavior, whereas others do not. A functional polymorphism in the gene encoding the neurotransmitter-metabolizing enzyme monoamine oxidase A (MAOA) was found to moderate the effect of maltreatment. Maltreated children with a genotype conferring high levels of MAOA expression were less likely to develop antisocial problems. These findings may partly explain why not all victims of maltreatment grow up to victimize others, and they provide epidemiological evidence that genotypes can moderate children's sensitivity to environmental insults."[206]

---

[203] Granick, J., "The Lie Behind Lie Detectors", *Wired* 2006:
http://archive.wired.com/politics/law/commentary/circuitcourt/2006/03/70411?currentPage=all
[204] "Study of youths and violence is shelved, but research critics are still upset", Knight Ridder Newspapers, May 21, 1993: http://articles.baltimoresun.com/1993-05-21/news/1993141073_1_violence-research-sullivan
[205] Ibid
[206] Caspi A1, McClay J, Moffitt TE, Mill J, Martin J, Craig IW, Taylor A, Poulton R., "Role of Genotype in the Cycle of Violence in Maltreated Children," *Science*, August 2, 2002: Vol. 297 no. 5582 pp. 851-854
http://www.ncbi.nlm.nih.gov/pubmed/12161658

This team found that a combination of an abused childhood and low activity of promoter levels for the monoamine oxidase-A (MAOA) gene resulted in high propensity for anti-social behavior. In previous studies, low levels of MAOA were correlated with violent behavior. Their study went further; it was performed with about 500 boys tracked from birth to adulthood and examined how levels of MAOA modulated the effects of childhood maltreatment. In the words of the authors, "These findings provide the strongest evidence to date suggesting that the MAOA gene influences vulnerability to environmental stress, and that this biological process can be initiated early in life."[207] These finding are summarized in Figure 5.2.

Figure 5.2 Link between the MAOA gene, childhood maltreatment, and antisocial behavior

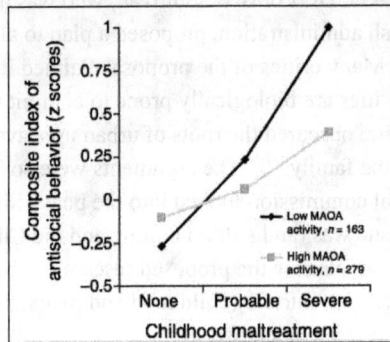

Source: "Role of genotype in the cycle of violence in maltreated children", *Science*, Vol. 297 no. 5582 pp. 851-854

The findings have been replicated in other studies, including meta analyses:

"Moreover, meta-analysis demonstrated that across studies, the association between maltreatment and mental health problems is significantly stronger in the group of males with the genotype conferring low vs. high *MAOA* activity. These findings provide the strongest evidence to date suggesting that the *MAOA* gene influences vulnerability to environmental stress, and that this biological process can be initiated early in life."[208]

A subsequent study of violent and non-violent offenders in Finland confirmed the importance of MAOA levels and added another possible genetic culprit.[209]

The authors of these and other studies that link genetics to behavior argue that genetics is not fate; while a genetic vulnerability may trigger subsequent biological and behavioral events,

---

[207] Ibid. and "The origins of violence. Nurturing Nature", *The Economist*, August 1, 2002.
http://www.economist.com/node/1259045
[208] Ibid.
[209] Tiihonen, J., "Genetic background of extreme violent behavior", Molecular Psychiatry, Oct 24, 2013:
http://www.nature.com/mp/journal/vaop/ncurrent/abs/mp2014130a.html

social support could protect even genetically vulnerable children from the negative consequences of maltreatment.[210]

5.9 Other possibly useful technologies

5.9.1 Tabulation of who is watching violent games, movies, and TV shows

There seems to be little doubt that watching violent media or playing violent video games can contribute to violent behavior. Helfgott makes the case by citing pertinent research; she says: [211]

> "More than 1,000 studies have been conducted on the effects of TV and film violence over the past 40 years, and in the last decade the *National Institute of Mental Health,* the *American Academy of Pediatrics,* the *American Academy of Child and Adolescent Psychiatry,* and the *American Medical Association* have reviewed these studies and concluded that TV violence leads to real-world violence (Senate Committee on the Judiciary, 1999). More recent research has further clarified the role media violence plays in creating real-world violence, suggesting that media violence produces short-term increases in aggression by triggering an automatic inclination toward imitation, enhancing autonomic arousal, priming existing cognitive scripts."[212]

However, uncertainties remain:

> Other researchers suggest that the attack on television violence by politicians, scientists, parents, and others is unwarranted and TV violence actually performs an innate human function by serving as "the most recent and least damaging venue for the routinized working out of innate aggressiveness and fear." [213]

5.9.2 Anticipate copy-cats

In 1915, the movie *Birth of a Nation* apparently inspired resurgence in KKK copycat lynchings and more recently (1999) the Columbine killings were followed by 400 related incidents.[214] [215]

---

[210] e.g. Kaufman J, Yang BZ, Douglas-Palumberi H, Houshyar S, Lipschitz D, Krystal JH, Gelernter J., "Social Supports And Serotonin Transporter Gene Moderate Depression In Maltreated Children," Proc Natl Acad Sc. Dec 7, 2004; http://www.ncbi.nlm.nih.gov/pubmed/15563601

[211] Helfgott, Jacqueline B. *Criminal Behavior: Theories, Typologies and Criminal Justice*, Sage, 2008, Chapter 10: "The Influence of Technology, Media, and Popular Culture on Criminal Behavior". http://www.sagepub.com/upm-data/19507_Chapter_10.pdf

[212] Fowles, Jib, "The Case for Television Violence," SAGE,: http://knowledge.sagepub.com/view/the-case-for-television-violence/n6.xml

[213] Ibid

[214] Coleman, L., "The Copycat Effect", 2004; http://www.lorencoleman.com/copycateffect/

In the film *Basketball Diaries* released in 1995, Leonardo DiCaprio's character has a dream in which he wears a long black trench coat and opens fire in a school with an automatic weapon and a shotgun. The Columbine shooters Dylan Klebold and Eric Harris are said to have been inspired by the movie. Klebold and Harris also played violent computer games such as Doom and Wolfenstein.[216]

Media coverage of the Columbine shooting also apparently inspired other copycat school shootings: Seung-Hui Cho, the gunman in the Virginia Tech massacre, referred to Columbine in video tapes he made of himself on the day of the shootings.[217]

Barry Loukaitis who was convicted of killing 3 fellow students in Moses Lake, Washington, in 1996 was said by prosecutors to have been inspired by the lyrics of a Pearl Jam video Jeremy[218], in which a student takes his own life in English class.

Grand Theft Auto Vicecity is a computer game that has been mentioned in a number of lawsuits as a factor involved in killings by young people. For example, Devin Moore was convicted of killing three members of the Alabama police force.[219] When captured, he is alleged to have said: "Life is like a video game. Everybody's got to die some time." His attorney called the game a murder simulator.[220] Moore was sentenced to death in 2005. [221]

Magnum Force is a 1973 film starring Clint Eastwood. A scene in the film shows a woman killed by pouring drain cleaner, which is basically lye, down her throat. This form of torture/murder was also used by the killers William Andrews and Dale Selby Pierre in what is known as the Hi-Fi murders in 1974 Ogden, Utah.[222] [223] The killers were executed.

---

[215] op cit; also see: Michael Atkinson, "The Movies Made Me Do It," http://www.villagevoice.com/1999-05-04/news/the-movies-made-me-do-it/
[216] Basketball Diaries - Columbine High School Massacre http://www.crsdrespect.org/columbine.html
[217] "Killer's Note: 'You Caused Me To Do This'", *ABC News*, April 17, 2007 http://abcnews.go.com/US/story?id=3048108
[218] Jeremy (song) http://en.wikipedia.org/wiki/Jeremy_%28song%29 and http://murderpedia.org/male.L/l/loukaitis-barry.htm
[219] "Devin Moore Sentenced to Death," October 7, 2005. http://www.tuscaloosanews.com/article/20051007/WVUA01/51006007
[220] "Can A Video Game Lead To Murder?" *60 Minutes* http://www.cbsnews.com/news/can-a-video-game-lead-to-murder-04-03-2005/
[221] "Can a Video Game Lead to Murder?" http://www.cbsnews.com/news/can-a-video-game-lead-to-murder-04-03-2005/
[222] Douglas, John E.; Olshaker, Mark (1999). *The Anatomy of Motive: The FBI's Legendary Mindhunter Explores the Key to Understanding and Catching Violent Criminals*, New York: Scribner. pp. 91–109. ISBN 978-0-684-84598-2.
[223] The "Hi Fi" Murders: http://murderpedia.org/male.A/a1/andrews-william.htm

The Tylenol Poisonings began in1982 when a 12-year-old Illinois girl who took one extra-strength Tylenol capsule collapsed and died at a hospital within hours. Three other people who lived near by died the same day. The Cook County Medical Examiner's Office found that all four had died of cyanide poisoning and "that the Tylenol capsules were filled with 65 milligrams of cyanide—10,000 times more than the amount needed to kill a person". Three more deaths followed. The product was promptly removed from the store shelves but many copycat murders have occurred since. "The FDA identified 270 incidents, 36 of which were labeled hard core true tamperings"[224]. In the two decades following there were copycat murders in New York and in Washington State involving several pain killers and some people were arrested, but to date the original Tylenol terrorist has never been caught. [225]

These cases (and many others in the literature) are at best anecdotal but they certainly illustrate the copy-cat phenomenon.

5.9.3 Control violent TV and game content

Hardly anyone who watches violent films or TV show or plays violent games becomes a terrorist, but some people, a very few, may. Viewing violent material may be a stop along the way for some who will become LWs or SIMADs. Conclusion: limit violent material in movies, TV shows and games and strengthen rating systems and "parental controls." Does this sound like censorship? Yes, but it could be warranted in an era of possible SIMAD.

There is precedent. CNN and other news media sources often self censor and do not show videos of material they judge to be too violent. Another instance: Timothy McVeigh who was convicted and executed for the killings and destruction in the Oklahoma City bombing of 1995, asked that national television stations be allowed to televise his execution. This was not permitted. Furthermore, Entertainment Network, an Internet company, sued for the right to webcast the execution (they intended to charge $1.95 per view and donate the funds to survivors of the bombing) and was denied that right in court.[226]

5.9.4 Collect frequency of visit data

Police and anti-terrorist officers often seek authorization to review hard drives to find the web sites that suspects have visited. But the question might be turned around and the proprietor of a

---

[224] Op. Cit. Helfgott, Jacqueline B.

[225] Decades later, the FBI sought a DNA sample from Ted Kaczynski apparently to check on his association with the poisonings. See: Schmadeke, S., Meisner J., and Gutowski, C., "Probe Of Tylenol Murders Shifts Focus To Unabomber's DNA", *Chicago Tribune*, May 19, 2011

[226] Mieszkowski K., and Standen, A., "The Execution Will Not Be Webcast," April 19, 2001. http://www.salon.com/2001/04/19/mcveigh_6/

web site might be asked who has visited their site. If the site presents violent scenarios, then it could be appropriate to ask if anyone has visited the site many times to find out who is watching perhaps compulsively. This will not be evidence in any event, nor will many proprietors be willing to share their client lists without judicial urging--and most importantly, it will limit personal freedom of the majority of people who use the site and are not potential LW or SIMADs, but "frequency of visit data" could be a brick in the building of a potential LW or SIMAD profile.

5.9.5 What do they read and write?

McVeigh is said to have read and been influenced by *Unintended Consequences*, a 1996 novel by John Ross, depicting government as a bully and revolt as the antidote, in essence, the use of arms in the defense of liberty (the cover shows an ATF agent shooting Lady Liberty).[227] McVeigh said that if he had read the book earlier, he might have used sniper attacks instead of bombing as his method of attack.[228]

Apparently, some potential Jihadists read *Islam for Dummies* and the *Koran for Dummies* reports the New Republic.[229]

Seung-Hui Cho, the Virginia Polytechnic Institute murderer, took a creative writing class at the Institute and the dialog in his plays so alarmed his professor that the professor alerted the police who, unfortunately said they were able to do nothing about the warning.[230]

In November 2002, Police Officer David Mobilio was shot and killed in a California gas station. Six days later postings on Internet read "Hello Everyone, my name's Andy. I killed a Police Officer in Red Bluff, California in a motion to bring attention to, and halt, the police-state tactics that have come to be used throughout our country. [.....]" The posting was signed by Andrew McRae, this was found to be an alias of Andrew Hampton Mickel.[231] Mickel was later apprehended and convicted of the murder.[232]

---

[227] Ross, J., *Unintended Consequences*, St. Paul, Mn: Accurate Press. 1996, p. 863. ISBN 978-1-888118-04-

[228] Michel, Lou and Herbeck, Dan. *American Terrorist*. p. 304. ISBN 0-06-039407-2

[229] Mehdi Hasan, "This Is What Wannabe Jihadists Order on Amazon Before Leaving for Syria," the New Republic, August 22, 2014. http://www.newrepublic.com/article/119182/jihadists-buy-islam-dummies-amazon

[230] Potter, N., et. al, "Killer's Note: You Caused Me to Do This," ABC News, Apr, 2007: http://abcnews.go.com/US/story?id=3048108

[231] Booth, W., "Murder Incorporated?" *The Washington Post*, April 4, 2005: http://www.washingtonpost.com/wp-dyn/articles/A24016-2005Apr3.html

[232] Andrew Hampton Mickel, http://murderpedia.org/male.M/m/mickel-andrew-hampton.htm

Joseph Andrew Stack III flew a general aviation airplane into a building in Austin, Texas on February 18, 2010.[233] Why? He wrote his manifesto telling what he was angry about (in this case angry at big business) and reading the manifesto might have provided warning that he was someone to be watched.[234] Theodore Kaczynski, Anders Behring Breivik, and others also wrote manifestos.

Many of these possible actions compromise our view of what is acceptable in a free society. But if a LW of the future is apprehended before he or she kills 100,000 or more people, the compromises to our freedoms could lie well beyond any of today's scenarios.

---

[233] Brick, M., "Man Crashes Plane Into Texas IRS Office," *New York Times*, February 18, 2010. http://www.nytimes.com/2010/02/19/us/19crash.html

[234] Wiesenthal, J., "The Insane Manifesto of Joseph Andrew Stack," http://www.businessinsider.com/joseph-andrew-stacks-insane-manifesto-2010-2

6. Legal and Ethical Aspects Related to Lone Wolf Phenomenon[235]

6.1 Introduction

Globalization, migration, growing inequalities, expansion of organized crime and terrorist groups, and increasing access of individuals to natural, technological and social resources, combined with outdated institutional, legal and governance systems have raised the world's vulnerabilities to new levels and are changing security paradigms. The diplomacy, foreign policy, military, and legal systems to address the new asymmetrical threats have yet to be set, as the UN, NATO and other security structures are based on nation-state as primary decision-making entity, which is becoming increasingly inadequate.

The UN Security Council Resolution 11580 Condemning Violent Extremism and to Prevent Travel, Support for Foreign Terrorist Fighters adopted in September 2014 is the first comprehensive international legal instrument that addresses the lone actor phenomenon. In addition to specifically calling upon all Member States to respect their obligations under international law to prevent the spread of radicalization and terrorism, it also asks for the development and implementation of "prosecution, rehabilitation and reintegration strategies for returning foreign terrorist fighters". [236]

Social frustrations, cyberspace, and easy access to new technologies provide the setting for a shift from terrorism planned and directed by centralized transnational extremist groups to leaderless terrorism [237] and attacks by small independent groups or individuals.

Improving capabilities to deal with lone actors has become an integral part of the highest priority strategies to reduce risks of terrorism, piracy, regional instability, and weapons and cyber attacks. However, while discourse about "lone actors"--or "lone wolves", or "solo terrorists"—increases, there is neither an academic nor an internationally-agreed-upon definition of the terms, as noted in the Introduction section of this book.[238] There are several reasons for this. While the

---

[235] Note: this chapter is drawn from an article that appeared in the NATO book published after the workshop "Lone Actors, an Emerging Security Threat" held in Jerusalem, November 3-6, 2014

[236] Security Council Unanimously Adopts Resolution Condemning Violent Extremism, Underscoring Need To Prevent Travel, Support For Foreign Terrorist Fighters, SC/11580, United Nations Security Council, New York City, New York, 24 September 2014 http://www.un.org/News/Press/docs//2014/sc11580.doc.htm

[237] Pluchinsky, D., *The Global Jihad: Leaderless Terrorism?* Wilson Center, Washington DC, June 20, 2006 http://www.wilsoncenter.org/event/blive-webcastb-the-global-jihad-leaderless-terrorism

[238] In the Introduction section, this book offers a tentative definition for a "lone wolf" terrorist, in the context of this study

---

attackers consider that they are fighters for some legitimate rights, those under attack see them as terrorists (another undefined term). In addition, it is also not always clear which degree of violence gets categorized as a lone wolf attack. Sometimes the distinction between criminal and "lone wolf" action is fuzzy—for example in school shootings--such as in 2012 by Adam Lanza in Newtown, Connecticut, or political assassinations--such as the crime committed by Yigal Amir in the 1995 assassination of Israel premier Yitzhak Rabin. The distinction between some lone wolves and martyrs (or suicide bombers or whatever the name) is even thinner. Some experts, such as Gabriel Weimann [239] state that there is no such thing as a "lone" wolf, that these lone actors, similar to wolves, are part of a pack. Nevertheless, it is generally accepted that "lone actors" operate by conviction, for a cause they believe fair, being driven by ideological, religious, ethnic, or social principles, and often--perhaps usually--have a mental disorder.

It is important though to point out that no matter what the definition, there are at least two types of "lone wolves": those who share the views of some radicalized organization (part of the leaderless terrorism)--such as alleged to be the case for Nidal Hasan, the U.S. Army psychiatrist who performed the Fort Hood mass shooting--and those who are not ideologically part of any specific group, who have their own principles or grievances for which they are ready to commit crimes in order to make the point--a good example is the Unabomber, Theodore Kaczynski, who mailed bombs to selected technologists over several years. Although there are similarities in the techniques they might use and the way they plot the attacks, generally, the underlying motivations of the two types are different. Therefore, the approaches—mostly the ethical and legal ones--to address the root causes or to reduce/annihilate the threats may be quite different.

Security and surveillance techniques and even the discourse about strategies used for lone wolves' understanding and eventual apprehension are mostly those used for terrorism. Further, generally, Europe continues to consider that lone wolf is only a derivative of terrorism. But the lone wolf phenomenon is more complex: the potential perpetrators are more difficult to detect and to defend against. Anders Breivik's horrific attacks were a wake-up call that put lone wolf seriously on the security agenda, recognizing the potential diversity and complexity of lone wolves and the distinction from other kinds of terrorists.

Lone wolf trends also differ considerably from those of more common terrorism. In some instances, there are links as to the perpetrators, but a comparison between the most targeted places shows significant variations. The Global Terrorist Attacks Database [240] shows that most terrorist attacks have been taking place in the Middle East and North Africa (MENA) and South

---

[239] Weimann, G., *Virtual Packs of Lone Wolves. How the internet made 'lone wolf' terrorism a misnomer*, Wilson Center, Washington, DC, 2014 https://medium.com/its-a-medium-world/virtual-packs-of-lone-wolves-17b12f8c455a

[240] Global Terrorist Attacks Database, START Project, University of Maryland, http://www.start.umd.edu/gtd/search/Results.aspx?charttype=line&chart=regions&casualties_type=&casualties_max=

Asia regions, while the most targeted regions of lone wolf attacks are North America (the U.S. accounting for 63% of all global lone wolf attacks between 1990 and 2013), Western Europe, Australia, and Russia.

Security and legal structures have to be updated to effectively address the increasing risks and destructive power of lone wolves, and to avoid the growing prospects of SIMAD. As shown in Chapter 3, future desktop molecular and pharmaceutical manufacturing, plus access (possibly via organized crime) to nuclear materials, could give single individuals the ability to make and use weapons of mass destruction. However, new technologies are also offering unprecedented possibilities for better security systems and pre-detection of dangerous individuals [see Chapter 3 and 5 for a description of some new and emerging capabilities.] However, their possible use is not covered by existing international legal frameworks and standards, raising the need to consider not only circumstances that are appropriate for their use, but their broader societal consequences as well.

Intelligence gathering and new security measures require enhanced tools. However, these techniques are not always desirable, nor do they necessarily meet the ideals of democracy and individual freedoms. Most lone wolf attacks have been carried out using guns; yet, in many countries, possession of guns is a constitutional right with slim chances of being changed. While cyberspace has become the fifth battlefield, it is also the backbone for free-flow of information, the very heart of a free-society; how much of its monitoring and eventual restriction is fair? How much is too much? Some argue that monitoring has to be regulated with clear indication of who has the right to monitor, how the information is used, where it is stored, and who has access to it. The irony is that people complain about government surveillance, while they voluntarily put their life online and apparently care even less about the more complete data collection by corporations for marketing and other purposes.

Public debate is necessary for citizens to understand the framework of the new threats and the required safeguards and security policies, the functions of science, and the changing influences in global politics and the positions of global actors. This would also help create a climate of trust and develop social relations that could help diminish the root causes of the threats themselves. Full engagement of the population and collaborative work of security, civil society, and government structures are sine qua non for successful new security regimes.

A critical point is to find methods to detect when a person should be considered a threat to society and how such cases should be treated [as described in Chapter 5 of this book.] But how accurate, legal, and ethical is profiling and the methods used for profiling? How should society treat individuals who are detected as potentially having genetic or mental problems that could turn them into an active threat to society? And what are the issues associated with false positives and false negatives? What are the roles and responsibilities of the public or the scientific community to detect and protect against potential SIMADs? These questions, as well as the right to weapons, have complex legal implications that have yet to be resolved at national—let alone international—levels. The eventually applicable legal or customary laws also depend on the type of lone wolf we are dealing with.

As one of the most important elements of the "glocolization"[241] of terrorism, the lone wolf phenomenon is expanding the scope and spectrum of counterterrorism to unprecedented levels. The occasional use of double standards in approaching non-state organizations or even certain nations could create confusion and mistrust among countries, allies, groups, and the public at large. Adapting the global security regime to the new power relations in international politics and different types of leadership in security is more than timely. Crises triggered by lone wolf actions built momentum that so far only translated into political capital and marketization of certain policies. Answering the question of what makes sense—legally, strategically, morally, and ethically--is very difficult, but crucial. And even so, as one participant in the RTD said, the dilemma remains: "who controls the controller?"

6.2 New Security Paradigms

As nation-states, we learned to live together much more peacefully, but as citizens, we still have a long way to go. Although traditional wars are becoming less frequent, according to the 2014 Global Peace Index[242], the world has become less peaceful every year since 2008. Similarly, the 2014 Fragile States Index[243] shows that out of the 178 countries rated for susceptibility to destabilization, 126 are in the alert or warning category. The 2014 Global Terrorism Index[244] indicates that terrorist activity is also on the rise: the number of people killed in terrorist attacks increased by 61% in 2013 compared to 2012, and the number of countries that experienced more than 50 deaths rose from 15 to 24 in this period. The rapidly changing political atmosphere and the increasing threat of "violent extremism and sectarian conflict, especially in fragile states" [245] top the security agenda.

A competitive dynamic appears to be at work among terrorist groups, leading to further escalation of cruelty and aggressive recruitment. Militant jihadists are promoting seventh-century

---

[241] Glocalization is a term that combines the words globalization and localization; http://encyclopedia.thefreedictionary.com/glocalization

[242] Global Peace Index 2014. Institute for Economics and Peace http://www.visionofhumanity.org/sites/default/files/2014%20Global%20Peace%20Index%20REPORT.pdf

[243] 2014 Fragile States Index. Fund for Peace http://ffp.statesindex.org/rankings-2014

[244] 2014 Global Terrorism Index; Measuring and Understanding the Impact of Terrorism. Institute for Economics and Peace http://economicsandpeace.org/research/iep-indices-data/global-terrorism-index

[245] Quadrennial Defense Review 2014, U.S.A. Department of Defense, Washington, DC, 2014 http://www.defense.gov/pubs/2014_Quadrennial_Defense_Review.pdf

ideology but are using modern tools—manipulating news media with videos of cruel behavior, using social media for communications, fund raising and recruitment, and employing modern weaponry. Their propaganda also appears to revitalize radicalization in some Western countries where it had been in decline, or at least under control.

Conventional defense and legal structures are not prepared to effectively address the new security environment. Current strategies mainly focus on security through the unique lens of counterterrorism and the "old narrative of West against terrorism" is not appropriate in the case of ISIS and might even ricochet, giving "fanatics justification to enter a holy war against democracies," as noted by José Manuel Barroso, President of the European Commission.[246]

The new asymmetrical threats do not respect sovereign borders or official systems, and cannot be addressed by any country or organization acting alone. Often, lone wolf acts involve activities and actors from several countries. Yet given the "dangerous sectarian, ethnic or tribal dimensions" of the current multiple crises, "many have seen sharp divisions within the international community itself over the response," remarks Secretary General Ban Ki-moon. [247]

If there is to be a new security paradigm, it should recognize that it is about contesting a philosophy. But when fighting a philosophy, there has to be another acceptable one to replace it, respecting complex cultural, religious, ideological and ethical aspects. Right now, we fight the philosophy that guides the lone actors—be it based on religious extremism or social discontent—but do little to offer an alternative, except rhetoric about a freedom that does not resonate with them. Thus, any new security paradigm requires innovative strategies that diminish the factors that favor the spread of threatening ideas. In short, security strategies should consider the ethical root-causes of new threats, as well as the ethics of responses in order to avoid aggravating circumstances. Sustainable security in a globalized world implies shared perceptions of socio-economic justice and security, as well as accountability. Even if trust is difficult to attain, cooperation based on common interest can be built.

Increased propaganda and invitation to participate in violence by extremist groups such as IS often appeal to people seeking a cause, with low self-esteem and few alternatives. Furthermore, reaction to the propaganda and proselyting opens the possibility that Muslims in the West are unfairly stigmatized, further aggravating homegrown terrorism threats. Why is it that people can be recruited? What in the western system of values makes them feel so worthless that they are

---

[246] José Manuel Barroso, President, European Commission, quoted in *Barroso Calls for More Decisive Commitment from the Arab World to Combat Extremists*, by Yann Zopf, World Economic Forum, Istanbul, September 29, 2014. http://www.weforum.org/news/barroso-calls-more-decisive-commitment-arab-world-combat-extremists

[247] Cited in: Charbonneau L, Nichols, M, *World leaders to gather at U.N. in shadow of Islamic State, Ebola crises*, United Nations, September 16, 2014 http://www.reuters.com/article/2014/09/21/us-un-assembly-idUSKBN0HG0W820140921

ready to die for causes that in many cases are not even their own? Combating the propaganda and the IS way of thinking, and the responses to criminal acts with increased menace may only amplify the intolerance and hopelessness that feeds extremism. Using the soft power of the media, supported by socio-economic policies that reinforce the counter-narrative of lone wolves might be one of the most efficient ways to reduce the phenomenon altogether.

6.3 Glocalization of Threats

Global extremist movements--mainly those linked to ISIS--are increasingly populated by people who have been recruited and radicalized online. Ideas that could have remained isolated before the Internet era now have the potential of triggering global movements. The materials range from online videos (including some that focus on abhorrent violence) to manifestos and real "do it yourself" manuals. The extremists' online strategy seems successful. An estimated 15,000 individuals from some 80 countries have joined groups like IS and the Jabhat al-Nusra.[248] Over 2,000 of these are from the EU and North America. The possibility that any of these adepts go back to their country of origin or another country to commit an attack as a lone wolf is of increasing concern. This was demonstrated by the January 7, 2015 Charlie Hebdo massacre in Paris, where at least one of the gunmen was trained in terrorist camps.[249]

Many Western countries are taking measures to stop the return of presumed terrorists and the debate has begun over the legal rights to prosecute citizens for fighting in another country's war or joining extremist groups abroad. These are dangerous measures that feed into the scope of the movement and could expand its spectrum, since those people will land somewhere, most probably in a country that has less security resources to deal with them. Meantime, at home, such measures might trigger even more alienation of those who could potentially share the extremists' cause.

Aggressive responses such as the airstrikes against militants in Syria and Iraq increase the risks of retaliation: "Whether [members of ISIS] carry out that threat by sending fighters back who they've trained, or whether they try to inspire lone wolves, or whether they simply captivate potential recruits and they go out on their own to carry out attacks," are possibilities Michael Chertoff, former Secretary of Homeland Security in the U.S., warns about. [250]

---

[248] Sink, J. "Americans radicalized in the Middle East back in United States", *The Hill*, September 22, 2014 http://thehill.com/policy/international/218494-white-house-radicalized-americans-back-in-us

[249] Jenkins, S., "Charlie Hebdo: Now is the time to uphold freedoms and not give in to fear", *The Guardian*, 7 January 2015. http://www.theguardian.com/commentisfree/2015/jan/07/charlie-hebdo-freedom-fear-terrorists-massacre-war?CMP=twt_gu

[250] Michael Chertoff in "Chertoff warns ISIS could hit US targets", interview by Molly K. Hooper, *The Hill*, September 30, 2014 http://thehill.com/policy/international/219267-chertoff-warns-isis-could-hit-us-targets

---

It is therefore important to understand the motivations of those people, to help their reintegration, to have adequate de-radicalization programs, and help them cope with potential post-traumatic stress disorder.

If not addressed effectively, over time, the problem will only expand, as globalization and regional demographic imbalances increase the rate of international migration. According to UNFPA, in 2010, some 214 million people (3% of world's population) lived outside their countries of origin. If current trends continue, by 2050, some 20% of Europe's total population might be Muslim[251], which could challenge Europe's generally secular values.

NATO needs new allies with younger populations. Southeast Asian strategists are concerned that the extremist ideology might attract significant number of the region's large Muslim population, mostly influenced by those returned from training camps. In Latin America, transnational organized crime might take advantage of the diffusion of extremism and use lone wolves to further destabilize governments and global security.

Meantime, the Internet is a medium for expressing independent ideas; these could also be radical ideas, which may become contagious. Before his atrocious act, Anders Breivik left tracks and shared know-how that is still online. His manifesto "2083: A European Declaration of Independence"[252] offers a manual for "Planning the operation". He instructs potential followers to "Do absolutely everything by yourself" and trust no one. However, there are still speculations about whether he was alone from ideological point of view. He might be a representative of lone wolf acting alone, but he claimed to have had contact with the English Defence League, the Norway Defence League (although that group denied it), and The Frankmasons (who expelled him after the attack). Yet, his ideology is still out there on the Internet, as are many others. The dilemma is how the medium can stay a place for expressing independent ideas and stimulate discussion about them, and yet avoid the uses that lead to expanding terrorism. There is precedence for removing objectionable material from the Internet (e.g. the closing of darknet child pornography sites); how can the line be drawn? What is the appropriate and acceptable compromise between unbounded freedom and security? And who should act and under what jurisdiction? These are questions that will be repeated often as the threats intensify.

In France, we already see the contest between freedom and control; a new decree that went into force in February 2015 allows the French government "to block websites accused of promoting

---

[251] Lorant, K., The demographic challenge in Europe, Brussels, April 2005.
http://www.europarl.europa.eu/inddem/docs/papers/The%20demographic%20challenge%20in%20Europe.pdf
although a recent Pew Research Center study estimates that Muslims will only represent 10% of the overall EU population: *The Future of World Religions: Population Growth Projections, 2010-2050*, Pew Research Center, http://www.pewforum.org/2015/04/02/religious-projections-2010-2050/
[252] Berwick, A., "2083: A European Declaration of Independence" on the website of the Federation of American Scientists (FAS) http://fas.org/programs/tap/_docs/2083_-_A_European_Declaration_of_Independence.pdf

---

terrorism and publishing child pornography, without seeking a court order." The new regulation asks Internet service providers (ISPs) to take down offending websites within 24 hours of receiving a government order. However, while the French Interior Minister Bernard Cazeneuve claims that the new decree is critical to combating terrorism, civil rights groups fear that it would give the government "dangerously broad powers to suppress free speech." [253]

As if to underscore this dilemma of contest between freedom and security, only 48 hours after the massive march for freedom of expression triggered by the Charlie Hebdo attack, France opened a criminal investigation of the comedian Dieudonné M'bala M'bala for his Facebook post: "Tonight, as far as I'm concerned, I feel like Charlie Coulibaly" – mocking the slogan "I am Charlie". His charges were "defending terrorism" (Coulibaly was the perpetrator of the Paris supermarket killings.) In reaction to the arrest, some media commented that "Expressing that opinion is evidently a crime in the Republic of Liberté, which prides itself on a line of 20th Century intellectuals – from Sartre and Genet to Foucault and Derrida – whose hallmark was leaving no orthodoxy or convention unmolested, no matter how sacred." [254]

Since a lone actor is a phenomenon, global legal framework, policies and international collaboration combined with local actions are required to counter the probability of attacks and to mitigate its possible effects.

6.4 Security and Trustfulness

Security and freedom is not a zero sum game—they go hand in hand and have to adapt to the complexity of new social and geopolitical developments, and technological advancements. If the security organizations are not a step ahead, malicious forces will find the openings. Also, since know-how is catching up fast, surveillance techniques have to evolve at an ever accelerating pace. Nevertheless, new tools have to be carefully disclosed and explained to the public and evolve in a legal framework to avoid distrust.

Trustfulness is of different types: among governments; among governments and the public; and among different groups. Nevertheless, the basic principles governing them are the same.

Data is the new transformative currency for all sectors and venues of life—good as well as bad--and affects security both positively and negatively. By now, all defense agencies and

---

[253] Toor, A., "France can now block suspected terrorism websites without a court order," *the Verge*, February 9, 2015. http://www.theverge.com/2015/2/9/8003907/france-terrorist-child-pornography-website-law-censorship
[254] Greenwald, G., "France Arrests a Comedian For His Facebook Comments, Showing the Sham of the West's "Free Speech" Celebration," *The Intercept*, Jan 14, 2015. https://firstlook.org/theintercept/2015/01/14/days-hosting-massive-free-speech-march-france-arrests-comedian-facebook-comments/

security organizations—national, regional, and international—have cybersecurity imbedded in their strategies as a top concern. Many lone wolves have been arrested worldwide, often before they were about to launch attacks. Many of these arrests were possible because the terrorists were tracked online. However, there is no international agreement over the extent of Internet freedom and the governments do not have clear legal frameworks under which to operate.

Many arrests have been based on police sting operations, and some people and defense attorneys have raised questions about whether these kinds of operations are entrapment, particularly when young and impressionable want-to-be lone wolves are involved. There is need also to articulate policy in this regime as well.

Surveillance partnerships are intrinsic to global security, but the level of surveillance acceptable between friends and allies is questioned, and even more so the cooperation with countries outside the "Five Eyes" or the enlarged UKUSA community, for example[255]. The Internet could only enable a more secure world with shared values if those values emerge collectively and are respected as such.

Joint exercises such as Unified Vision 2014 (UV14) that took place in May 2014 show a great deal of cooperation and trustfulness among those countries as well as the agencies involved. However, it is certain that there were many eyes watching, trying to find gray-zones or back doors for compromising such actions if they come up in real life or even just to know the capabilities. These are challenging but inviting grounds for lone wolves and lone wolves' herd leaders.

Edward Snowden has been accused of compromising secure NSA data; this act and others like it raise suspicions and distrust among allies and play into the hands of those with malicious intent. There is no consensus about whether Snowden's act was treason or heroism. If it was a crime, does he qualify as lone wolf? Nevertheless, the Snowden case has triggered deep reconsideration about the collection, protection, and use of big data.

The quiet cooperation of the government with the IT sector has been shaken, with most big companies announcing that they are sealing up cracks in their systems to restore users' trust, which is their "currency". The IT and telecommunications companies set off a new arms race, denying requests to provide data not required by existing laws, or use the "gray areas" for cooperation, while also making it more difficult for security and intelligence agencies around the world to access their systems. This might seriously impede the efforts to identify potential lone wolves; "sooner or later there will be some intelligence failure and people will wonder why the intelligence agencies were not able to protect the nation," warned Robert Litt, General Counsel

---

[255] "Five eyes" refers to an alliance between Britain, the US, Australia, New Zealand and Canada

of the Office of the U.S. Director of National Intelligence.[256] The irony is that generally, the security sector only piggybacks on the surveillance infrastructure for massive data collection developed by the business sector (including Facebook, Google, email services and chat rooms, and marketing databases), which lobbies for larger surveillance capacity itself.

The "spy on bad guys but not good guys" sounds like a double standard and is ill defined, but such simplifications can improve public understanding and support for measures that are meant to improve public security. The reality and perception should be that the average user is in greater danger from Chinese or other hackers than the security agencies. Closing data sources to security agencies might deny them a valuable resource, and yet leave them open to lone wolves. As mentioned in Chapter 4 The Cyber Dimension, huge data breaches have already taken place; the JPMorgan Chase incident affected some 83 million households and businesses, in summer 2014. The 2014 attack on eBay netted 233 million personal records; the Montana Health Department reported that more than a million records had been tapped. And some of the other companies that suffered attacks in 2014 include: Target, Neiman Marcus, Michael's, Yahoo, AT&T, PF Chang, UPS, Home Depot, Google, Apple iCloud, Goodwill Industries, and Dairy Queen. [257]

The question remains: will the sophistication and acceleration of new intrusion techniques be so fast that debates about surveillance technologies become obsolete even before new safeguards can be implemented? Can security agencies stay ahead of the wave? And when security gets to be outsourced to artificial intelligence, how will multiple choice decisions system provide a moral framework, compassion, and empathy?

6.5 Legal Framework and Legitimacy

In the case of lone actors, the popular quote by Gnaeus Pompeius Magnus (Pompey) is quite relevant: "Stop quoting laws, we carry weapons." [258]

Right now, there is no internationally-agreed upon legal framework and terminology to address the lone wolf phenomena. Therefore, surveillance, reporting and eventual prosecution are difficult, as "illegality" might differ from country to country. This is so even in the context of

---

[256] Sanger D., E., Perlroth, N., "Internet Giants Erect Barriers to Spy Agencies", *The New York Times*, June 6, 2014
http://www.nytimes.com/2014/06/07/technology/internet-giants-erect-barriers-to-spy-agencies.html
[257] Walters, R., "Cyber Attacks on U.S. Companies in 2014," The Heritage Group,
http://www.heritage.org/research/reports/2014/10/cyber-attacks-on-us-companies-in-2014
[258] Favourite Historical Quotes, Roman Army Talk, (pg 2), Retrieved October 2, 2014
http://www.romanarmytalk.com/7-off-topic/152763-favourite-historical-quotes.html

collaborating countries, such as members of the OSCE, thus limiting the effectiveness of cross-border co-operation between law enforcement and related organizations.[259]

The main difficulty is the one mentioned before, that there is no internationally-agreed-upon definition of lone actor or lone wolf. Generally, it is accepted that lone wolves commit an act of violence with significant impact and that they operate outside of a command structure. However, in the U.S. legal context, which added a "Lone Wolf" Amendment to the Foreign Intelligence Surveillance Act in 2004, someone can be considered lone wolf, on simple basis of suspicion, without necessarily requiring "probable cause." [260]

Another important gap is that the international security and customary law systems are tailored to address inter-state disputes. The Geneva Convention should be updated to cover intrastate conflicts, terrorism, and potential lone wolf activities, and the support for the International Criminal Court should be global and enforced. So far, the international legal instruments that could be applied to lone actors are generally those related to terrorism; therefore, they cover only one specific lone wolf type. Creating an effective legal framework is difficult because of changes in political powers, the complexity of new asymmetric security threats, and the diversity of stakeholders and organizations that are involved.

The UN Security Council Resolution 11580 Condemning Violent Extremism and to Prevent Travel, Support for Foreign Terrorist Fighters adopted in September 2014 is the first comprehensive international legal instrument that addresses lone actors. However, experts warn that enforcement is virtually impossible, since it is really up to each country to implement its provisions and design appropriate counter measures. Also, the resolution does not authorize military action by any country.

UN Resolution 1540, adopted in 2004 is meant to prevent non-State actors from acquiring and using WMD. Although, as UN Secretary General, Ban Ki-moon noted, "there are no right hands for these wrong weapons."[261] After ten years, nearly 90% of Member States had taken measures to implement the requirements of the resolution. However, enforcement mechanisms for WMD regulations are yet to be adequate. While the chemical and nuclear treaties have enforcement mechanisms, the deadlock for the Biological Weapons Convention continues; in the meantime, the developments in the field and the related threats continue to evolve [see Chapter 3].

---

[259] OSCE Online Expert Forum Series on Terrorist Use of the Internet: Threats, Responses and Potential Future Endeavours. Final Report. OSCE, 2013 http://www.osce.org/secretariat/102266?download=true

[260] Intelligence Reform and Terrorism Prevention Act of 2004: "Lone Wolf" Amendment to the Foreign Intelligence Surveillance Act. CRS Report for Congress, December 29, 2004

[261] *Ban calls for intensified efforts to remove threat of weapons of mass destruction*, UN News Centre, 28 April 2014 http://www.un.org/apps/news/story.asp?NewsID=47668

The Arms Trade Treaty entered into force on December 24, 2014, setting the first global standards for the transfer of weapons and support efforts to prevent their use for genocide and crimes against humanity. However, the treaty only applies to conventional weapons and combat systems.

The legal framework covering terrorism comprises some 14 universal legal instruments and several amendments and relevant UN Security Council Resolutions. [262] While these instruments create a relatively acceptable--but patchwork--framework for action, implementation and enforcement are extremely difficult; the UNODC Terrorism Prevention Branch provides mainly technical assistance to Member States in designing their national legal frameworks. Nevertheless, UNODC addresses a large scope of interests and a broad spectrum of concerns, including identifying areas that could help extremist organizations, such as transnational organized crime, money laundering, and corruption. The Comprehensive Convention on International Terrorism would expand the present framework with increased obligations of Member States for cooperation—including prevention procedures, as well as extradition—but as of February 2015, negotiations were deadlocked.

The "United Nations Global Counter-Terrorism Strategy" [263] requires "development and social inclusion agendas in order to reduce marginalization and the subsequent victimization that propels extremism and terrorist recruitments", and respect for human rights. However, the 2014 report of the Director of the Counter-Terrorism Implementation Task Force Office (CTITF) [264] notes that "shortcomings continue to hamper national efforts", ranging from the "absence of the basic legal framework for the Security Council sanctions resolution requirements" to confusion over existing sanctions regimes, as well as "ineffective liaison with the private sector."

The lack of international agreement over who is or is not considered a terrorist adds confusion and difficulty to the application of international law. The UN list of terrorist organizations is much less inclusive than that of the NATO countries (which is not unanimously-agreed upon, either). The UN does not include groups like Hezbollah, which the U.S., Canada, and the UK consider a terrorist group. Recently, the PKK had a twist of fate, with several EU governments and the U.S. arming it to help fight the IS. Over the years, there were several organizations that had such twist of fate that then backfired, causing increased security concerns and fueling distrust.

---

[262] United Nations Action to Counter Terrorism; International Legal Instruments
http://www.un.org/en/terrorism/instruments.shtml
[263] UN United Nations Global Counter-Terrorism Strategy: activities of the United Nations system in implementing the Strategy http://www.un.org/en/ga/search/view_doc.asp?symbol=A/66/762
[264] Statement of Mr. Jehangir Khan Director CTITF Office Closing Remarks at Greentree, 14 May 2014,
http://www.un.org/en/terrorism/ctitf/pdfs/Closing%20Statement%20of%20Mr%20Khan%20at%20Greentree.pdf

Although cyberspace is the most important tool for recruiting and training potential lone wolves and has the potential for causing large scale disasters, there is no international agreement on the future governance of the Internet. The UN resolution 'Right to Privacy in the Digital Age' adopted by the UN General Assembly in December 2013 calls upon all States to "review their procedures, practices and legislation regarding the surveillance of communications, their interception and the collection of personal data, including mass surveillance, interception and collection, with a view to upholding the right to privacy by ensuring the full and effective implementation of all their obligations under international human rights law."[265] This is also echoed by the NETmundial Multistakeholder Statement issued at the end of the Global Multistakeholder Meeting on the Future of Internet Governance, held in Brazil, in April 2014. There is no agreement on the process for further negotiations. Some countries, including China and Russia want the negotiations to take place within the UN framework, while others, such as the U.S., Australia and several European nations favor passing the responsibility to a group that is not dominated by governments.

Many eyes are on the debate for amendments to the Electronic Communications Privacy Act in the U.S., for eventual emulation. As of this writing, the debate continues[266], since both the companies and the civil liberties organizations are fighting back, because the new draft does not address some of their concerns that were included in the first draft. Similar debate is going on in Canada, over Bill C-51--Anti-terrorism Act, which would enact and make amendments to several security regulations and the Criminal Code.[267] Civil society organizations, the corporate sector and the opposition parties[268] warn that if the bill passes, some "17 government agencies and even foreign governments" will also have access to citizens' "sensitive private information."[269]

Although more than 70 countries have or are developing drones and other devices for remote-control warfare, there are no international laws regulating their use, let alone enforcement mechanisms. As with other new technologies, the use of these devices is two-faceted—it can help in defense and reconnaissance, but could also become a tool of destruction if used by an enemy, including lone actors. In fact, according to the FBI, Rezwan Ferdaus, a Lone Wolf, intended to use a radio controlled model airplane as a drone to deliver his bombs in an attack on

---

[265] The right to privacy in the digital age. Resolution adopted by the General Assembly on 18 December 2013 68/167. http://www.un.org/ga/search/view_doc.asp?symbol=A/RES/68/167

[266] S.356 - Electronic Communications Privacy Act Amendments Act of 2015 https://www.congress.gov/bill/114th-congress/senate-bill/356

[267] House Government Bill C-51 Anti-terrorism Act, 2015
http://www.parl.gc.ca/LegisInfo/BillDetails.aspx?Language=E&Mode=1&billId=6842344

[268] Elizabeth May Speech on Bill C-51, April 24, 2015. http://elizabethmaymp.ca/elizabeth-may-speech-on-bill-c-51/

[269] Tell the government to stop Bill C-51 before it's too late. https://stopc51.ca/

the Pentagon.[270] And in January 2015, a "hobby" remotely controlled quad-copter drone crashed on the White House lawn.[271] The high-end drones cost about $1,000; many are made in China and can carry a high-resolution video camera payload.

International cooperation and consistency are of utmost importance for a successful strategy to reduce extremism worldwide. Regional and local participation are required for legitimacy of Western involvements.[272] "[W]hen we cannot explain our efforts clearly and publicly, we face terrorist propaganda and international suspicion; we erode legitimacy with our partners and our people; and we reduce accountability in our own government," notes President Obama. [273]

The same principle applies internationally. Actions have to be based on trust and ethics. However, the present legal framework and even the ethical principles have yet to be determined. While some cases are clear-cut and can be prosecuted without ambiguity, in many potential LW cases, serious dilemmas exist and the dialogue continues. One such dilemma is concerning the publication of scientific research or material that could be exploited or used by potential LW. As seen in Chapter 1, the participants in the RTD study were divided over the answer; and so is the scientific and legal community in general.

The media are crucial to increasing awareness among citizens about the real extent of threats. Citizens who are informed about the potential consequences of lone actors would demand policymakers to increase efforts to create a relevant global legal and security regime, and would be more cooperative in addressing the potential underlying factors of lone wolf terrorism. Present policy frameworks that set the stage for national and international policies are often linked to some financial aspects; taxpayers should be aware that the costs of lone wolf impacts could sky rocket over time, if the phenomenon is not adequately addressed. Countermeasures introduced now could be much less expensive than those required if action is delayed.

6.6 Instead of Conclusions

Using different potential LW-related situations and measures for avoiding LW threats, the following table offers a summary of the current legal and ethical status and likely response actions.

---

[270] FBI, "Man Sentenced in Boston for Planning Attack on Pentagon", November 1, 2012.
http://www.fbi.gov/boston/press-releases/2012/man-sentenced-in-boston-for-plotting-attack-on-pentagon-and-u.s.-capitol-and-attempting-to-provide-detonation-devices-to-terrorists
[271] Schmidt., S. M., and Shear, D. M., "A Drone, Too Small For Radar To Detect, Rattles The White House, "*New York Times*, January 26, 2015: http://www.nytimes.com/2015/01/27/us/white-house-drone.html?_r=0
[272] "ISIS and the End of the Middle East as We Know It", Wilson Center, October 09, 2014,
http://www.wilsoncenter.org/event/isis-and-the-end-the-middle-east-we-know-it
[273] President Obama "America Must Always Lead": President Obama Addresses West Point Graduates
http://www.whitehouse.gov/blog/2014/05/28/america-must-always-lead-president-obama-addresses-west-point-graduates

Table 6.1 Current Legal and Ethical Status concerning potential LW Situations

Assumed situation	Current Legal Framework Adequacy	Challenge to Today's Ethics	Possible responses
Person identified having stolen or illegally possessing nuclear materials	Yes	Low	Prosecution
Person identified illegally possessing bomb-making materials	Yes	Low	Prosecution
Person has purchased significant quantities of potentially hazardous materials	Mostly yes	Medium	Eventual prosecution Increased enforcement or new laws restricting sale of critical materials
Online discussions or published material to recruit terrorists	Not	High	Nothing in calm environment. In a paranoiac environment, freedom of speech restrictions, mass screening programs, and likely prosecution.
Scientific paper submitted for publication could help LW to make more effective weapons	Not	High	Publisher's and peer judgment. In a more paranoiac environment, external, special review panels and potential withholding of publication.
Person has published a paper on Internet on how to make a new infectious virus	Not	High	Nothing in a calm environment. In a paranoiac environment, forced to take it offline, denial of publication.
A person, previously identified as a LW or SIMAD threat is apprehended in the process of building a biological weapon capable of killing millions	Not	High	1. Confidentiality privileges denied 2. Mental assessment and eventual institutionalization 3. Potential prosecution 4. Screening of their network 5. New laws restricting sale of critical materials
Mass screening programs identify 3,000 persons or more who seem to have LW or SIMAD proclivities	Not	Very high	1. Grading their level of threat 2. Heated debate over potential false negatives or positives 3. Monitoring, eventual treatment 4. Denied certain privileges 5. Screening of suspects' networks

The big dilemma remains about how to reconcile the different systems of values: Can civilization ever achieve a peaceful level of coexistence without losing diversity? This theme alone can be the subject of an entire book and years of jurisprudence debates.

7. Conclusions and Recommendations

7.1 Introduction

What do we think we know? Lone Wolves have been with us for many years and probably will be for many to come. We know that LW attacks happen with great frequency and it seems are a nightly feature on TV news and other media. We know that the attention that LWs receive can influence susceptible persons to perform similar copycat acts or even acts they view as more "heroic." We know that the weapons of the LW have the potential for expanding in capability, particularly in biotechnology and cyber crime. We know that security technology is improving as well but will never be 100% effective in pre-detecting precision and therefore potentially harmful persons who are not identified (false negatives) will be a perennial threat. We know also that no matter how excellent identification techniques become, some people will be falsely identified and therefore unwarranted stigma will also be created (false positives). We believe that LWs may be able to kill and injure many more people in a single episode in the future than in the past. We also believe that when (or if) the threat of mass casualties becomes more tangible, compromises with civil liberties may become more acceptable (for example, routine scanning of emails, purchases, and sites visited may be seen as necessary to obtain a reasonable sense of security).

In the following paragraphs we summarize important conclusions that flow from the chapters of this book and present recommendations for policies and actions that may help reduce the LW and SIMAD threats. These recommendations may well duplicate activities that are already underway but are not in public view.

7.2 The RTD survey

Some important conclusions about future LW and SIMAD threats may be drawn from the RTD study, including:

Many important questions examined in this study remain unanswered.

The panel of experts that participated in the RTD came to no consensus and sometimes sharply disagreed about issues such as:
- The effectiveness of techniques for detecting LWs and SIMADs
- The potential of profiling for detecting prospective LWs or SIMADs

- The possibility that LW terrorism would expand into SIMAD, and if it were to occur, its timing and scope
- Means of coping with people who are suspected of having SIMAD potential
- The strategies and policies to diminish LW threats
- The appropriate compromise between civil liberties and improved security
- The right to publish scientific papers that contain potentially threatening "how to do it" weapon information

The size of a potential SIMAD attack may increase exponentially with time.

The panelists were asked for their judgments about the likely timing and size of an initial SIMAD attack. Plotting their average answers as a semi log graph results in a straight line; this implies a constant rate of growth in the number of people killed in a single SIMAD attack of 12% per year, reaching 100,000 by 2067. However, there was considerable disagreement and widespread differences among the responses. Another interpretation of the RTD data is that if there has not been a SIMAD episode by 2050, its probability will diminish greatly.

There was no consensus about the timing of a SIMAD attack.

According to the participants in this study, a SIMAD attack may never happen or may happen sooner rather than later. According to non-security people, it might happen soon, but almost all those who placed the timing of the attack beyond 2075 or never, identified themselves as experts in "Security." Is this perception biased by over-confidence in current security strategies?

LW and SIMAD targets are very diverse.

Some panelists suggested that in addition to population at large or certain population segments or infrastructure, targets could also include agriculture and the environment.

S&T represent both threat and opportunity

New S&T developments were mentioned as potential sources for new, largely unexpected weapons to be used by LWs. The panelists mentioned several, including biological (viruses, alien species), as well as some low tech, which may have weapons implications: new effective poisons distributed in novel ways, invasive species as a bioweapon, and dispersible nanotech agents.

However, new S&T discoveries can also be used for potential detection of LWs (e.g. fMRI, genetics) and prevention of potential attacks (e.g. nano-swarms of small detectors for sensing the presence of toxic agents.) New technologies might help detect propensity to violence in individuals and should be pursued; these technologies include physical and psychological scanning, brain imaging, psychological and behavioral profiling, and perhaps genetic analysis.

Cyberspace is both a threat and opportunity.

The computer can be a weapon, even a weapon of mass destruction if it is used to disable systems that are required for society to function. The RTD participants mentioned some possible targets. However, cyber systems also offer the potential for detection and prevention—by monitoring communication, social media, purchases, and so on.

Security strategies to curb LW attacks and their consequences should include prevention as well as resilience.

The participants in the RTD study suggested three categories of measures designed to identify or influence persons who might otherwise become SIMADs:
- "Soft approaches" (e.g. public awareness campaigns, emphasis on family, social justice, and ethics, and educational reform)
- "Hard approaches" (e.g., genetic screening, fMRI brain scanning, psychological conditioning)
- "Vigilant monitoring" of persons thought to be predisposed to violence.

A comprehensive worldwide database is needed.

We recommend that a database be constructed containing complete descriptions of past LW events--both completed and thwarted as well as environmental scanning and assessment of trends, relevant S&T developments, points of potential access to WMD or information about how to construct them, and best anti-LW practices. Access to sensitive portions of this database should be tightly controlled. Data should be included to help identify attractive LW and SIMAD targets so that those places can be given priority in planning security measures. A panelist in the RTD study suggested that such criteria might be used to form a prospective "hit list" to be used in estimating the risks associated with particular venues.

Restricting civil liberties as part of strategies to prevent LW and SIMAD attacks is a controversial matter

Similar to public debates, the participants in the RTD study were split over the question of whether or not intrusion into privacy was justified: on the one hand defending the need for intrusion based on the potential for mass casualties and on the other, opposing it, to protect privacy.

Addressing this threat requires a complex and long-time continuous effort and cooperation of national and international authorities

The means and the targets of LWs are multiple and complex; even more so are the strategies to address them. These strategies should include international agreements and cooperative programs of data and intelligence sharing, as well as policing efforts. Cooperative efforts are of course already underway, but they are scattered at national or regional levels, or restricted to certain domains. The threats of SIMAD imply that the scope and spectrum of these efforts have to expand (ideally globally) and may continue for decades. They imply consideration of economic, social, moral, political, and educational activities.

Religious leaders have an important role to play in reducing LW and SIMAD threats

Religious incentives were identified by most participants in the RTD study as the motivation behind future LW attacks. When it becomes clear that LW can morph to SIMAD and many thousands could be killed in a single event, will leaders of some religious movements still tolerate or encourage terror attacks in the supposed support of their ideologies and teachings?

LW and SIMAD targets

The RTD panel overwhelmingly selected North America as the likely place for future SIMAD and LW attacks. One panelist, however, felt that Asia was more likely to be the initial target. He used these criteria to designate the location:

[The SIMAD attack would likely be in a place] *with large population and geographic base, conflict is significant. Access to means could be high. Visibility can be moderate to high. Security systems to prevent may be porous. Belief systems are strong and polarizing...*

Media Issues

The media help shape perceptions about the nature, severity, and risks of LW attacks. Some RTD respondents suggested that LWs might even consider media coverage as a criterion in planning their attacks. We recommend that the media be encouraged to provide a balanced rather than a sensationalized picture, avoiding aggrandizement of "successful" attacks and attackers while responsibly reporting on successful interdiction. The media also should be encouraged to follow up by reporting on penalties that convicted Lone Wolves have received.

Media attention and region of the participants might have created a potential bias in the RTD conclusions.

No study is ever completely free of biases and we conclude that the answers provided by the RTD panel might have been influenced by several factors: the relative lack of participants from Asia, Africa, and Latin America; the concentration of recent attacks in North America and their exposure in the media; and as in all opinion-based studies, the product is very dependent on the particular set of participants. Another panel might have produced different answers.

Methodological aspects

The high level of confidence that the RTD participants expressed in their answers surprised us. The answers to the questions they were addressing were intrinsically unknown, for the most part unknowable, or at best, obscure, yet self-confidence, even in the face of disagreement among them was high. From a methodological standpoint the meaning of this high confidence is unclear. However we are sure that high confidence should certainly not be confused with high accuracy in assessing the probability of these events.

More research and informed action is essential to avoid potential future catastrophes

The central conclusion emerging from the RTD study is that the threat of the LW and SIMAD will increase over time and minimizing them should be a high priority on the security agenda around the world. Further research and analysis focusing on LW psychology and decision making (as opposed to terrorism in general) could help in developing new approaches based on improved understanding of the phenomenon and effective modes of its defeat or containment.

7.3 New Technologies

In the hands of malicious actors, some emerging technologies can pose new and significant threats. Constructing policies that limit access to these technologies by unauthorized persons should be high on the agenda. Designing such measures however requires consideration of trade-offs between issues of security, human rights, and freedom of publication.

This dilemma was also studied by the FESTOS EU project that, in its reports, included some policy principles for applying control of materials and knowledge developed in research and development[274]. These principles, modified somewhat to apply to our case, are summarized as follows:[275]

> Research and development projects carry the obligation to understand unintended uses that may be made of the developments. It is the researcher and developer that bear this obligation and we recommend that all advanced projects consider the possibility of unintended uses and their consequences, and if threatening, means for reducing their risks.
>
> Oversight procedures at institutions developing technologies that may have security implications need to be reviewed and strengthened if necessary In particular, procedures need to be in place to prevent theft of materials and information from both outside and inside research and development institutions as well as in manufacturing and storage.
>
> Inventories of dangerous materials must be meticulous and up to date.
>
> Proposals to publish information resulting from research and development of potentially dangerous products, methods, and materials should be reviewed in a system that oversees and approves publications that may include such information.

---

[274] Op cit. FESTOS Report
[275] Op cit. FESTOS Report

Proposals for research and development projects should include clear descriptions of methods for meeting appropriate security and safety standards. Security measures should be considered in the initial design of products and research plans. This process is known as "Security by Design."

The significance of this issue requires that it be considered at a national level. One suggestion is for a National Science Security Council to establish overall science security policies for the nation.

Creation of a Security Impact Statement (similar to an Environmental Impact Statement) should be considered by planners and funding agencies to encourage systematic thinking about potential security threats, impacts, and methods of protection.

Security R&D

New and extended research and development programs for improving security should be considered to reduce new opportunities for mayhem that will open to lone wolves in the future. Examples include:

- Detection and identification of future materials and hidden nano-systems in various environments
- Mitigating the threat of hacking into the Internet-of-Things
- Security implication of nano-products, nano particles and nano materials
- Technologies that will be able to detect non-metal weapons
- Detection and identification of new and unknown bio agents in the lab and in the outside environment
- Automatic alarm systems that warn of bio terror attacks
- Identification and defense against swarms of miniature robots
- Monitoring ongoing developments in fields such as brain research, brain-scanning and brain-computer interfaces, including psychological aspects, with special attention to potential abuse and for potential use in early detection of LWs
- Enhancing research on advanced techniques for timely identification of people with malicious intentions

We recommend that this objective receive priority in programs of funding agencies worldwide and in relevant national and international programs. We recommend that development of new technologies be accompanied by a security assessment process to lower the possibilities of malicious use from the outset.

Foresight and future scenarios

Foresight studies should be implemented periodically and horizon scanning should be implemented continuously to assess new security needs and responses. Scenario building and

---

identification of signals of change should be included. Such research will help increase preparedness and reduce surprises.

7.4 The Cyber Dimension

When the book of lone wolf history is finally written decades from now, it may be cyber attacks that have proven to be the most damaging of all. While most people today may equate the idea of cyber attacks with credit card or identity theft, the true situation is potentially far more serious. Our research and that of others who have asked about the possible futures of cyber terror, indicates that there are two principal domains: active and passive and for both domains there are offensive and defensive possibilities. Some examples:

- Active offensive: Stuxnet[276], and the attack on Sony Pictures[277]
- Active defensive: "Hacktivists," called Anonymous, who, like vigilantes of old, shut down terror recruitment sites[278] and claim that they have closed 800 Twitter and other social media accounts[279]
- Passive offensive: advertising campaigns to deter terrorist recruitment
- Passive defensive: development of advanced firewalls and encryption techniques

While cyber terror activities are most often software-based and rely on information networks, hardware approaches are also involved and include, for example, the possibility of miniature eavesdropping devices, of essentially undetectable firmware implanted in hard drives that allows surreptitious readouts, of "Trojan Horses" that deliver spy code imbedded in ordinary-appearing programs, and of "back doors" that provide easy access to those who know about these hidden routes. Several hardware and firmware approaches were mentioned in Chapter 5.

Cyber terror activities can be conducted by a single lone wolf terrorist or by a group of programmers or, if hardware or firmware is involved, electronic experts. The work can be self-inspired or performed for state sponsors or terrorist groups such as ISIS who mean to inflict economic and political chaos, or gain notoriety. This turbulent mix is growing in its scope of targets and means of accomplishment, difficulty of detection, and severity of impacts. Cyber operations are now embodied in military-like organizations charged with accomplishing national

---

[276] Op cit. David Kushner
[277] Continuing coverage of the hacking of Sony Pictures Entertainment, *NBC News* online: http://www.nbcnews.com/storyline/sony-hack
[278] "#Anonymous vs #ISIS: the ongoing skirmishes of #OpISIS", The Cryptosphere, March 20, 2015. http://thecryptosphere.com/2015/03/20/anonymous-vs-isis-the-ongoing-skirmishes-of-opisis/
[279] Petroff, A., "Hundreds of social media accounts apparently linked to terrorist group ISIS have been shut down", CNNMoney (London) February 10, 2015. http://money.cnn.com/2015/02/10/technology/anonymous-isis-hack-twitter/

objectives and national defence, freelancers, spies and counterspies, and now with vigilantes as well.

The conclusions and recommendations that flow from our research are:

Potential Cyber Attack Targets

Targets abound and new technologies and applications will broaden the opportunities to attack in ways about which we can only speculate. Imagine, for example, the possibility of a terrorist posing as just a passenger who brought his laptop aboard, hacking into the instrumentation or control system of aircraft in flight.[280] Or consider how attractive the Internet-of-Things will be to an amateur or freelance professional or state sponsored hacker; or the healthcare networks that are being formed; or the automated international financial clearing system that transfers over 38 trillion of dollars annually[281]; or credit card transactions; or stock markets transactions; and the list of possible targets is much longer. All offer the opportunity for mayhem, shakedowns, competitive advantage, profit, blackmail, and notoriety for the perpetrator.

We recommend that agencies that have the responsibility to anticipate and thwart cyber attacks perform a series of simulations in which they act as though they were potential SIMADs. The objectives of these sessions would be not only to identify targets likely to be high on the lists of the malintents, but to find criteria that they might use for target selection. This would help law enforcement authorities identify high priority targets to harden and protect. Some are obvious, but others may not be.

Persons of Interest

On the TV news magazine *60 Minutes*, Steve Croft interviewed Jon Miller, an ex-hacker who is now Vice President of Cylance, an antivirus software manufacturer about the attack on Sony Pictures. When Steve Croft asked how many people in the world could pull this off, Miller said, "There are probably three, four, five thousand people that could do that attack today."[282] Since that's a relatively small number, we recommend that such a list of talented people be compiled and updated periodically, including samples of their code or hacks they are thought to have performed so that forensic analysis in the future might identify telltale signatures in code that point to the originator, as might be done with handwriting or linguistic analysis, or with greater precision, fingerprinting.

---

[280] "Update: Hacker On A Plane: FBI Seizes Researcher's Gear", *The Security Ledger*, April 17 2015.
https://securityledger.com/2015/04/hacker-on-a-plane-fbi-seizes-researchers-gear/
[281] "Payment, clearing and settlement systems in the United States", CPSS – Red Book – 2012.
https://www.bis.org/cpmi/publ/d105_us.pdf
[282] "The Attack on Sony," Interview by Steve Kroft, *60 Minutes*, April 12, 2015

This would be practical even if the list included many more that 5,000 people. Of course, not everyone who has written code for hacking is likely to be a terrorist; so, judicial review and oversight is in order.

Bragging Rights

If not already implemented, a continuous review of social media and other sites for self-revealing tweets and statements about intent should be maintained since many LW like to boast about their intended actions, motives, and targets. Computer hackers in particular often boast about their prowess. If possible, this review ought to be automated using intelligent programs of the sort that the IBM Watson computer could mount, that is, it could perform analysis by inferences drawn from the sites under review. It will be no easy task to implement this across all public and hidden social media, but it could be very productive.

Cyber Terrorism as a Weapon of Mass Destruction
As pointed out by the RTD respondents and many others, cyber systems can be weapons of mass destruction. One of the respondents said:

> Cyber terror has been defined as: the intimidation of civilian enterprise through the use of high technology to bring about political, religious, or ideological aims, actions that result in disabling or deleting critical infrastructure data or information and/or resulting in massive loss of life.[283]

We will inevitably raise cyber barriers around crucial systems and infrastructure. It is reasonable to expect that terrorists, either singly or in concert, will attempt to defeat these defences. Therefore, we recommend that all cyber defence systems have the ability to sound a figurative alarm when attacked and if possible, to build in means for tracing the origin of the attacks.

Encryption

Recommendation: There should be no organization in the world with greater capability in encryption or de-encryption than a selected agency involved in anticipation or apprehension of LW and SIMAD terrorists. Perhaps this is already the case.

7.5 Detection: identification of people with malicious intent

Our review of possible pre-detection technologies leads to several conclusions. Although some promising pre-detection techniques are in hand or are on the horizon, none will be perfect and

---

[283] With permission: Dr. William Tafoya, Professor and Director of the University of New Haven National Security Program

therefore these methods will inevitably falsely identify some people as threats and allow others who are really threats, to remain undetected. Furthermore, some of the possible techniques start us on a slippery slope that leads to significant compromises to freedoms we enjoy and ultimately to Big Brother totalitarianism. Watchdogs and whistle-blowers need to be encouraged and safeguards enforced. But note: the worse the perceived threat, the higher the incentive for action, and the lower will be our guard.

Several conclusions follow:

Minimize false reports by using multiple methods based on different principles

Minimizing false reports should be a key objective in the development of detection systems used to identify potentially threatening individuals. In addition, all personnel who use such systems should be aware of the accuracy limits of their systems. No matter what technique is employed, single pre-detection methods will probably be inadequate and perhaps unfair; corroboration will be required. To help improve accuracy, multiple methods based on different operating principles should be used. The methods should synergize, one strong where the other is weak.

Beware of witch-hunts

There is little doubt that third party reporting ("see something, say something") has been and will be effective in identifying potential lone wolves and SIMADs. This has already proven to be the case. In the limit, well-meaning people constitute an enormous force for pre-detection since they represent "crowdsourced" eyes and ears. But the question is how to limit its possible abuse. As noted in Chapter 5, we are reminded of the era of McCarthy and the Salem witch hunts when hysteria, paranoia, and fear led to false and ill-informed accusations. Certainly, public reporting should be encouraged, but the design of such systems should include systemic safeguards against abuse.

Include Screening and Far-out-Methods in the R&D Agenda

Effective mass screening would require the development and proof of psychological tools not yet at hand. Mass screening is likely to be extremely unpopular, no matter how intense the threat it is designed to detect. Nevertheless, some forms of screening occur in the normal flow of life and these sources may yield some clues about future behaviour; for example, psychological tests associated with enlisting in the armed forces, in screening for security clearances or police forces, or for other forms of employment. An effort should be initiated to determine what "ordinary" screening forms might be used for this purpose--perhaps with modest changes if appropriate, and if these purposes are divulged to the subject.

The use of fMRI, electroencephalography, and genetic analysis to reliably identify potential lone wolves before they act will be impractical for many years to come. But as pointed out in

---

Chapter 3, some of the technologies, when perfected, may help identify individuals with a propensity for violence. Yet these tools, even when developed to their full potential over the next decade or so, are not likely to serve as the basis for mass screens or to produce unequivocal evidence of malintent.

The Sum of All Measures

Because no single system seems to hold the key to "pre-crime" detection, we envision the evolution of a layered system in which warning signs from many sources are combined to identify individuals who seem to have a higher chance than others of exhibiting violent behavior in the future.

Many pre-detection programs are in development; emphasis should be placed on those that promise to be effective, near term, and less intrusive on human rights. One approach that seems to qualify is the development of software for automated screening of big, diverse databases. Such software would tap different databases to identify specific individuals by cross-referencing printed words, photographs, and videos, tracking purchases of critical materials, reviewing arrest records, scanning self-published information about intent, "reading" manifestos that incite to terror, reviewing MAOA levels in the brain if available, and following communications among known malintents. An automated system of this sort might be used to form a composite risk number associated with an individual; the higher the number, the more carefully the person would be monitored. This system would require judicial oversight but finding individuals that appear in multiple terror related situations could lead to an effective system for "vigilant monitoring."

If this approach were to be used, measures would be required to minimize false accusations and assure privacy-- a tough prescription.

7.6 Legal and Ethical Aspects

The global and national legal frameworks and security regimes should be adapted to the new types of threats and international political and security power relations. An important element is addressing the increasing threat and potentially destructive power of lone wolves and to avoid the growing prospects of SIMAD.

The UN Security Council Resolution 11580 Condemning Violent Extremism and to Prevent Travel, Support for Foreign Terrorist Fighters adopted in September 2014 is the first comprehensive international legal instrument that addresses the lone wolf phenomenon. It calls upon all Member States to prevent the spread of radicalization and terrorism and to adopt strategies for the reintegration of returning foreign terrorist fighters. Nonetheless, it does not have enforcement mechanisms and it is up to each country how to implements its provisions.

Since the lone wolf is a glocalized phenomenon, global policies and international collaboration combined with local actions are required to counter the probability of attacks and to mitigate the possible effects. However, sensitive ethical implications and complex global relations challenge the negotiations and potential changes both at national and international levels.

One of the greatest impediments is the lack of internationally-agreed upon concepts and definition of actions and inconsistency in reactions. For example, the distinction between criminal and "lone wolf" action sometimes is fuzzy (let alone the "martyr" interpretation). Therefore, clear international classifications—as much as possible--could help produce effective legal frameworks and policies.

Global security requires partnerships among nations, but the level of surveillance acceptable between friends and allies is questioned, and even more so the cooperation among the global community as a whole. There should be an international agreement concerning the rights to prosecute citizens for fighting in other countries' conflicts or joining extremist groups abroad.

Intelligence gathering and new security measures require advanced tools and methods. Research should be expedited to find methods to detect when a person should be considered a threat to society and how such cases should be treated. However, these techniques might not always meet the ideals of democracy and individual freedoms.

The Internet could only enable a more secure world with shared values if those values emerge collectively and are respected as such. We need to seek a global consensus on how to deal with Internet and other media that help promote terrorism. Many lone wolves have been arrested worldwide, often before they were about to launch attacks. Many of these arrests were possible because the terrorists were tracked online. However, there is no international agreement over the extent of Internet freedom and the governments do not have clear legal frameworks under which to operate. What is the appropriate and acceptable compromise between unbounded freedom and security? How much is too much?

Similarly, protective measures should be established globally to control access to and trade in technologies that might have value to terrorists. There is a need for adequate enforcement mechanisms for all types of WMD, as well as drones and other devices for remote-control warfare, and other new technologies. While some of these technologies can help in defense and reconnaissance, they could also become powerful tools of destruction.

Addressing the new security paradigm has to consider offering an alternative philosophy to replace the philosophy that leads to terrorism in general and lone wolf specifically. It is important to understand the motivations of those people, to help their reintegration, to have adequate de-radicalization programs, and help them cope with potential post-traumatic stress disorder. The alternative should respect complex cultural, religious, ideological and ethical aspects of the new social and international reality. Using the soft power of the media, supported by socio-economic policies that reinforce the counter-narrative of "recruiters" might be one of the most efficient ways to reduce the lone wolf phenomenon altogether.

7.7 Concluding Remarks

We believe that a new kind of arms race is developing. On one hand, is the possibility of increasingly destructive weapons falling into the wrong hands, and on the other, the development of new methods of surveillance and pinpointing individuals with malintent. Will the methods of detection be adequate and timely enough to avoid catastrophe? It is often said that one of the major purposes of futures research is to provide lead-time to decision makers and the decision process; we have lead-time. We hope it will be used to reduce the threat we see in front of us.

Albert Einstein is supposed to have said, "The world will not be destroyed by those who do evil, but by those who watch and do nothing." We hope we have done something with this book; that's the reason we wrote it.

Appendices

Appendix A. The RTD Questionnaire

Global Opinion Studies
Polls, Surveys, Crowdsourcing, and Real Time Delphis

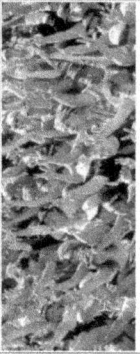

Prospects for Lone Wolf Terrorism
Opportunities and Policies to Minimize the Threat

To see a reference paper on this topic _click here_

Please answer the questions in the form below. When you return please enter as a returning participant and use this email address: email **and this study code:** SIMAD. **This study is scheduled to close on 2013-12-15.**

By pass introduction

Introduction

World events make it clear that *lone wolf* terrorism is a potent and growing worldwide threat. Defined as single individuals performing or intending to perform acts of violence, lone wolf terrorism is insidious because it is so hard to detect before the fact. A few of the most well known examples are:

Anders Breivik, in Norway, who bombed government buildings in Oslo and killed eight people and then went on to kill 69 teenagers on the island of Utvya, in 2011.

Nidal Hasan, who opened fire at a military base in the United States, in 2009.

Ted Kaczynski, the Unabomber, who despised modern technology and mailed bombs to the people who he thought were threatening, between 1978 and 1995.

Timothy McVeigh, the bomber of the Oklahoma City federal building who killed 168 people and injured over 600, in 1995.

Nabil Ahmad Jaoura, shot tourists at the Roman Amphitheatre in Amman, Jordan killing one and injuring six in 2006

Mohammed Merah, in Toulouse and Montauban, France killed seven people including three children in 2012

The list is much longer, of course, and includes episodes involving ricin, anthrax, and sniper and incendiary attacks and other forms of mayhem.

We define *lone wolf* terrorists as single individuals acting essentially alone who kill or injure people or inflict significant damage on essential infrastructure at a single instant or over time, or plan to do so, in order to bring about political, religious, or ideological aims. Cyber terror is also included when it results in massive destruction or loss of life. Lone wolf terrorists are not directed by outside hierarchies. Targets may be specific groups or undifferentiated masses of people; motivations may range from religious imperatives to perceived injustices, or to psychoses. Weapons may include firearms, homemade bombs, computers, poisons and biological agents, and in the future, easier to produce weapons of mass destruction or disruption (WMD) including self manufactured chemical weapons and manufactured biological agents, as well as weapons of military grade that may have been taken from military reservoirs or laboratories, or purchased on the black market.

The name given to lone wolves who use or plan to use WMD is SIMAD. This term was coined in a scenario prepared by the Millennium Project in 2002, as part of the study on future science and technology management strategies. It stands for Single Massively Destructive and is meant to convey the threat posed by psychopathic individuals who plan to do harm with weapons that are capable of killing and injuring large numbers of people or destroying or incapacitating infrastructure. In the 11 years that have passed since the scenario was written, SIMAD has become an even more chilling real possibility.

The lone wolf problem is global and multifaceted: on the one hand is the possibility of escalation of weaponry including genetic coding and essentially undetectable computer viruses. On the other hand, there may be new means for early detection of potentially aberrant behavior. Some scientists argue that fMRI and other imaging tools could help to identify potential terrorists. The social implications of searching for a potential lone wolf also needs study. If a threatening person could be discovered before the fact, what would society do about it? So far, all answers point toward a restriction of civil liberties.

Freedom to publish also enters the picture. On May 2, 2012, a scientific paper was published in *Nature* after a controversy on whether the processes described in the paper could have harmful consequences. It described how H5N1 avian virus flu could be modified into a form that would speed up human contagion. There was an uproar over its possible publication but it is now on the street.

This questionnaire is part of a study designed to:

Reach a broader understanding of the nature and potential for escalation of the lone wolf threat

Identify technology domains that have the potential to increase or change the nature of this threat

Explore policies and approaches to limiting access to information about these technologies

Explore policies and approaches to counter these technologies if they are deployed

Identify plausible means for diagnosing potentially threatening individuals

Consider humane and socially acceptable means for dealing with these individuals

Examine the consequences of policies designed to reduce lone wolf threats in the long-term, such as education

Identify possible early warning techniques

Distribute findings of the study to interested and affected groups including legal, scientific, law enforcement, civil liberty, first responders, and political agencies

These objectives will be accomplished through analysis of pertinent published literature, performing this Real-Time Delphi study involving experts from around the world who represent the many skills and expertise required by the topic, synthesizing their judgments and analyzing and publishing the findings of the study. We will be particularly alert to problems of conflicting goals and needs, for example, the possible trade-offs between public security and individual liberties. We ask all participants to avoid providing any information that might be considered classified.

Instructions

The questionnaire is in the form of a matrix. The rows list potential developments, variables, and policies; the columns pose questions about the items that appear on the rows. Each cell in the questionnaire has several entries: A place for you to enter your response to the question. The current group responses will appear after you have entered your estimate and after a specified number of responses has been received. The number of responses received so far will also appear. The link "REASONS CLICK HERE" appears at the bottom of each cell. This takes you to a new page where you will be asked for the reasons behind your answers. Please read the comments that others have submitted and respond if you wish. When you return now or later in the study you can change your inputs based on the comments.

There are two ways to enter your answers on this page. 1) You may enter your answers one at a time by pressing the GO button in each cell or 2) you may press the SUBMIT THIS PAGE button where ever it appears to enter your answers all at once. In either case, your answers will be entered immediately, and the form will return to your screen, containing your answers and the group answers that will have been updated to include your estimate. You may change your answers anytime, if you wish. If not otherwise stated, the questions deal with the world as a whole rather than a specific country, and if there is ambiguity, please assume the question deals with the global situation. Finally, at the bottom of this page you will also find a link to a new page where you may submit comments, suggestions, and additions. Please follow this link; your inputs are important. We really appreciate your participation.

This study is scheduled to close on 2013-12-15 and you may make changes in your answers until then.

Please return to the questionnaire often. When you return please enter as a returning participant and use this email address: email and use this and this study code: SIMAD.

Please remember to press SUBMIT at end of questionnaire.

Questionnaire

Questions	Question	Confidence
What percentage of terrorist attacks do you think will be carried out by lone wolves in 2015? (details) *In US: 1935-1977= 7% (Source: Mark Hamm, Indiana State University)* *1978-1999= 26% (Source: Mark Hamm, Indiana State University)* click here for example click here for example	Please enter a number without a percent sign or punctuation Submit only this cell: go Reasons for your answer click here	What is your confidence in your answer? ○ Very high ○ High ○ Middle ○ Low ○ Very low Submit only this cell: go Reasons for your answer click here
By what year will a lone wolf terrorist kill & injure the number of people shown here?	Please enter your estimate of the year of occurrence in any or all of the boxes. (Please answer 2100 if you think it will never happen.) 500 1,000 5,000 10,000 100,000 1,000,000 Submit only this cell: go Reasons for your answer click here	What is your confidence in your answer? ○ Very high ○ High ○ Middle ○ Low ○ Very low Submit only this cell: go Reasons for your answer click here
When do you think a lone wolf terrorist may first try to use a weapon of mass destruction? (details) click here for example click here for example Submit this page	By what year is there a 50/50 chance of occurring? (Please answer 2100 if you think it will never happen.) Submit only this cell: go Reasons for your answer click here	What is your confidence in your answer? ○ Very high ○ High ○ Middle ○ Low ○ Very low Submit only this cell: go Reasons for your answer click here

4

Where do you think an actual lone wolf terrorist attack of the sort described in Question 3 might first occur?

(details)

click here for example
click here for example
click here for example

○ North America
○ South America
○ Asia (ex. China)
○ China
○ Africa
○ Europe
○ Middle East

Submit only this cell go
Reasons for your answer click here

What is your confidence in your answer?

○ Very high
○ High
○ Middle
○ Low
○ Very low

Submit only this cell go
Reasons for your answer click here

5

However distorted the thinking, what do you think is the single most important motivation that will fuel lone wolf attacks over the next decade?

(details)

click here for example
click here for example
click here for example

○ Religious incentives
○ Installing new forms of government
○ Seeking a place in history
○ Redress of perceived wrongs
○ Insanity
○ Raising money and conscripts
○ Other (Please describe in reasons)

Submit only this cell go
Reasons for your answer click here

What is your confidence in your answer?

○ Very high
○ High
○ Middle
○ Low
○ Very low

Submit only this cell go
Reasons for your answer click here

6

What do you think the primary target of lone wolf attacks will be over the next decade?

(details)

click here for example

Submit this page

○ Population at large
○ Specific population segments
○ Agriculture
○ Infrastructure
○ Government officials
○ Other (Please describe in reasons)

Submit only this cell go
Reasons for your answer click here

What is your confidence in your answer?

○ Very high
○ High
○ Middle
○ Low
○ Very low

Submit only this cell go
Reasons for your answer click here

7

Do you think serious attempts to search for lone wolf terrorists who are capable of carrying out an attack using a weapon of mass destruction will be made before such an attack occurs?

(details)

click here for example

click here for example

○ Yes
○ No

Submit only this cell go

Reasons for your answer click here

What is your confidence in your answer?

○ Very high
○ High
○ Middle
○ Low
○ Very low

Submit only this cell go

Reasons for your answer click here

8

What technology is likely to be most effective for the detection of people with evil intentions? Consider fields such as psychology, brain imaging, observation of unusual behavior, etc.)

click here for example

click here for example

click here for example

○ Mass psychological screening
○ Monitoring of purchases of critical materials
○ Monitoring communications and social media
○ Third party reports of unusual behavior
○ Brain physiology
○ Genetic screening
○ Other (Please describe in reasons)

Submit only this cell go

Reasons for your answer click here

What is your confidence in your answer?

○ Very high
○ High
○ Middle
○ Low
○ Very low

Submit only this cell go

Reasons for your answer click here

9

How successful do you think the search strategies of Question 8 will be? In a realistic future, of 100 possible lone wolf attacks how many are likely to be avoided?

click here for example

click here for example

Submit this page

Please enter a number between 0 and 100 (do not use a per cent sign)

[ ]

Submit only this cell go

Reasons for your answer click here

What is your confidence in your answer?

○ Very high
○ High
○ Middle
○ Low
○ Very low

Submit only this cell go

Reasons for your answer click here

10

Do you believe that scientific and technological papers and other publications that contain information potentially useful to terrorists should be controlled or withheld?

click here for example

Please give us your thinking in reasons.

- ○ Yes
- ○ No

Submit only this cell go

Reasons for your answer click here

What is your confidence in your answer?

- ○ Very high
- ○ High
- ○ Middle
- ○ Low
- ○ Very low

Submit only this cell go

Reasons for your answer click here

11

Assume that some lone wolf terrorists choose to use massively destructive or disruptive weapons; from what fields might these weapons come?

click here for example
click here for example
click here for example

- ○ Nanotechnology
- ○ Biotech and synthetic biology
- ○ Nuclear physics
- ○ Computers/Communications
- ○ Power generation and transmission
- ○ Agriculture and food
- ○ Other (Please describe in reasons)

Submit only this cell go

Reasons for your answer click here

What is your confidence in your answer?

- ○ Very high
- ○ High
- ○ Middle
- ○ Low
- ○ Very low

Submit only this cell go

Reasons for your answer click here

12

Are "soft" approaches such as education reform or public awareness campaigns likely to be effective in dealing with the lone wolf threat in the long term?

(details)

click here for example

Submit this page

Please provide your thinking in reasons.

- ○ Yes
- ○ No

Submit only this cell go

Reasons for your answer click here

What is your confidence in your answer?

- ○ Very high
- ○ High
- ○ Middle
- ○ Low
- ○ Very low

Submit only this cell go

Reasons for your answer click here

#	Question		What is your confidence in your answer?
13	*Are actions which intrude on privacy of people or otherwise compromise their civil rights justifiable in view of the threats?*	○ Yes ○ No *Submit only this cell* [go] *Reasons for your answer* click here	○ Very high ○ High ○ Middle ○ Low ○ Very low *Submit only this cell* [go] Reasons for your answer click here
14	*If people are identified as potential lone wolves of the sort we have already seen currently, how do you think society should deal with them?* *(details)*	*To proceed to the answer form please* click here	
15	*What if they have been engaged in building weapons that could produce mass destruction or disruption?* *(details)*	*To proceed to the answer form please* click here	
16	*What steps, if any, do you think should be taken to minimize lone wolf threats? When?*	*To proceed to the answer form please* click here	
17	*What steps, if any, do you think should be taken to minimize threats that may achieve destruction on the scale depicted in Question 2? When?*	*To proceed to the answer form please* click here	

[Submit this page]

[Submit this page]

ADDITIONAL SUGGESTIONS

To submit comments and suggestions click here

Appendix B. RTD Responses—full listing

Question 1: What percentage of terrorist attacks do you think will be carried out by lone wolves in 2015?

Question 2: By what year will a lone wolf terrorist kill & injure the number of people shown here?

Question 3: When do you think a lone wolf terrorist may first try to use a weapon of mass destruction?

Question 4: Where do you think an actual lone wolf terrorist attack of the sort described in Question 3 might first occur?

Question 5: However distorted the thinking, what do you think is the single most important motivation that will fuel lone wolf attacks over the next decade?

Question 6: What do you think the primary target of lone wolf attacks will be over the next decade?

Question 7: Do you think serious attempts to search for lone wolf terrorists who are capable of carrying out an attack using a weapon of mass destruction will be made before such an attack occurs?

Question 8: What technology is likely to be most effective for the detection of people with evil intentions? Consider fields such as psychology, brain imaging, observation of unusual behavior, etc.)

Question 9: How successful do you think the search strategies of Question 8 will be? In a realistic future, of 100 possible lone wolf attacks how many are likely to be avoided?

Question 10: Do you believe that scientific and technological papers and other publications that contain information potentially useful to terrorists should be controlled or withheld?

Question 11: Assume that some lone wolf terrorists choose to use massively destructive or disruptive weapons; from what fields might these weapons come?

Question 12: Are "soft" approaches such as education reform or public awareness campaigns likely to be effective in dealing with the lone wolf threat in the long term?

Question 13: Are actions which intrude on privacy of people or otherwise compromise their civil rights justifiable in view of the threats?

Question 14: If people are identified as potential lone wolves of the sort we have already seen currently, how do you think society should deal with them?

Question 15: What if they have been engaged in building weapons that could produce mass destruction or disruption?

Question 16: What steps, if any, do you think should be taken to minimize lone wolf threats? When?

Question 17: What steps, if any, do you think should be taken to minimize threats that may achieve destruction on the scale depicted in Question 2? When?

Other Comments Made by the Respondents

Question 1: What percentage of terrorist attacks do you think will be carried out by lone wolves in 2015?

Number of answers: 48
Average response: 25.6%, over a range that extended from 3% to 65%.

Question	Min	Max	Average	Median
1. What percentage of terrorist attacks do you think will be carried out by lone wolves in 2015?	0.5	65.0	24.6	25.0

Distribution:

Range of answers	Number of Responses
0-10	12
10.1-20	5
20.1-30	20
30.1-40	5
>40.1	6

Confidence in the answers was fairly high:

Confidence	Number of Responses
Very High	1
High	16
Middle	21
Low	6
Very Low	4

In the analysis of responses to this question, the judgments by participants about confidence were plotted against their judgments of anticipated percent of terror attacks that would be conducted by lone wolves. We were interested in whether or not there was a correlation between the stated confidence and the substantive answers. The individual answers of the 50 or so people who contributed to this question are plotted below.[285] Each point represents the two answers that a

---

[285] Since many responses overlapped (for example two respondents may have said they thought that 30% of terrorist attacks would be by lone wolves and each had the same confidence in their answer) the graph would not show that concentration. So a small random "jitter" was applied to confidence values to separate them in this space. This "jitter" was also applied to all similar graphs in the remainder of this report.

---

single individual provided: one on confidence and the other on percent of terror attacks by lone wolves.

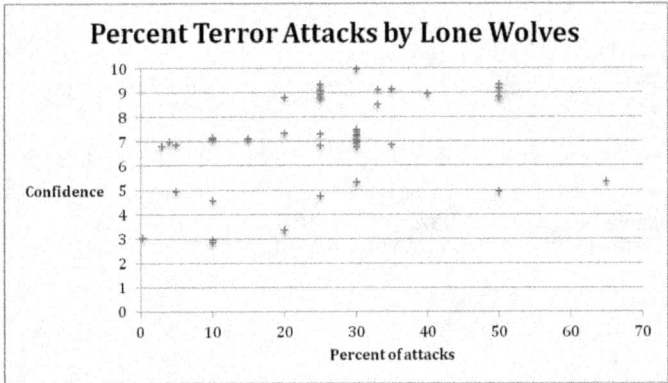

There appears to be some correlation between confidence and perception of threat of attacks by lone wolves: the higher the judgment about likelihood of attacks, the greater the respondents' confidence in their answers.

Text responses:

I think many will be confused by the definition as to whether it does include people who act by themselves but belong to an ideology. In the end the increase in danger is not as such lone wolfs but the increasing destruction potential that can be carried out by one person. I do think that aspect should be added to the survey and the background should be divided between the loner and the individual linked to a group.

availability of information for the design and development of WMD, open sites for targets-shopping areas, malls, sport events, agricultural sites-available news media and camera on smart phones to cover events, NEXT generation WMD-invasive species bioweapons, DIY biotech to develop new bioweapons, suicidal mentality of lone wolves,

Today number of groups involved is becoming target of security forces thus SIMAD attacks will increase. More so these cannot be tracked before execution nor can be traced back to any organization.

Lone wolf terrorist is minor to my opinion.

Terrorism usually is organized by groups or countries.

Lone Wolves can be self recruited, or ideologically recruited by groups, but acting on own initiative.

The threat of lone wolf attacks has increased, but not significantly.

I took into consideration people acting alone, not linked to some groups/organizations.

I suspect that ATTEMPTED attacks by individuals will be higher than my 25% estimate. However, individual attacks seem to fail more frequently.

Los patrones de conducta que tienen los SIMAD, generalmente son derivados de crisis familiares internas en la infancia o adolescencia propiciados por carencias económicas, religiosas o de inestabilidad social de sus padres, lo que provoca resentimientos contra la sociedad y gobierno, bajo la óptica de que continuarán presentándose crisis económicas y guerras religiosas en el futuro, al incrementarse la población, proporcionalmente se incrementa el porcentaje de SIMAD.

[Google Translate of this response: The patterns that have SIMAD are generally derived from internal family crises in childhood or adolescence brought about by economic, religious or social instability, parenting deficiencies, causing resentments against society and government, from the perspective of that they continue to present economic crisis and religious wars in the future, with increasing population, the percentage of SIMAD proportionally increases.]

Boston Marathon Bombing 4/15/2013 carried out by Dzhokhar & Tamerlan Tsarnaev is a more current example of the kind of terrorist acts I expect to be perpetrated in the near term future.

Question 2: By what year will a lone wolf terrorist kill & injure the number of people shown here?

The respondents were asked to judge the year when a lone wolf terrorist attack might kill 500; 1,000; 5,000; 10,000; 100,000; and 1 million people.

The dichotomy was quite sharp: half the group believed that such an attack might occur before 2050 and half the group, by 2100 or never. There were no responses for the period 2050-2075.

Period	Number of respondents
prior to 2015	2
2015-203	6
2030-20150	8
2050-2075	0
after 2075 or never	15

There was considerable disagreement among respondents. For example while average year in which 100,000 deaths might occur due to a SIMAD attack was judged to be 2067, the median year was 2100 or later[286] and the distribution of judgments was bimodal.

Question	Min	Max	Average	Median
2A. By what year will a lone wolf terrorist kill & injure 500 people?	2000	>2100	2024	2016
2B. By what year will a lone wolf terrorist kill & injure 1,000 people?	2000	>2100	2031	2020
2C. By what year will a lone wolf terrorist kill & injure 5,000 people?	2001	>2100	2047	2027
2D. By what year will a lone wolf terrorist kill & injure 10,000 people?	2017	>2100	2058	2030
2E. By what year will a lone wolf terrorist kill & injure 100,000 people?	2017	>2100	2067	2100
2F. By what year will a lone wolf terrorist kill & injure 1,000,000 people?	2017	>2100	2080	2100

Once again, confidence was fairly high:

Confidence	Number of Responses
Very High	4
High	5
Middle	19
Low	7
Very Low	7

---

[286] The questionnaire instructions asked for a response of 2100 if the participants felt the correct answer was "never."

Text responses:

Until now we didn't see a terror event with more then some hundred casualties even not 9/11. High numbers can be achieved by a LW only with a nuclear bomb which isn't accessible to individuals. A bio epidemic will be stopped quickly.

A Ph.D. dissertation at the University of Maryland, making use of the [conventional] Delphi Technique [Olaf Helmer was a member of that dissertation committee], completed in 1986, forecast, among other things, that by 1995 political terrorism in the U.S. would increase by more than 50 percent over the 1984 rate. Subsequent analysis evidences spikes in the temporal trend data, but also a steady increase over time.

There are many reasons to think that LW and SIMADS particularly, will pursue actions that affect or kill many people. Political motivations, for example, will benefit from panic and media attention; both increase as the numbers people affected or killed increase.

Development of techniques and tools and knowledge [is] expanding--e.g. DIY biotech, synthetic biology; decrease in personal restraint of violent behavior globally, extensive training of violence from Internet, former military, reactions of poor and oppressed used as motivations, lack of spiritual growth in Y-generation and millenials to lead to less self-restraint and hence gravitate to violence.

Question 3: When do you think a lone wolf terrorist may first try to use a weapon of mass destruction?

Responses received: 45
Average of the dates provided: 2033

About half the group thought that lone wolf terrorist might attempt to use weapons of mass destruction in the interval between 2020 and 2040:

Range of answers	Number of responses
Prior to 2015	4
2015-2020	9
2020-2040	21
2040-2060	6
2060-2099	0
2100 or Never	4

Confidence in the answers was relatively high:

Confidence	Number of Responses
Very high	2
High	11
Middle	16
Low	8
Very low	2

Text responses:

Anthrax was already used by a LW using envelopes in the US. Can happen every day since then but with low number of casualties.

[numerical answer: 2014] *Consider including Cyber Terrorism as a WMD. Cyber terror has been defined as: the intimidation of civilian enterprise through the use of high technology to bring about political, religious, or ideological aims, actions that result in disabling or deleting critical infrastructure data or information" (Tafoya 2011). To this definition the following *should* have been included: "... or information AND/OR RESULTING IN MASSIVE LOSS OF LIFE."*

[numerical answer: 2015] *availability of biological weapons even made from scratch exists presently; use of invasive species as biological weapons exists presently, nuclear weapons and radioactive materials to make radiological bomb-presently exists; chemicals to make chemical weapons presently exists, Internet information to make EMP pulse weapons presently exists; capability to make such weapons airborne with balloons or put into confined population areas such as sports arenas, malls, or public political events presently exists. Cyberweapons to knock out power grids or damage critical infrastructure presently exists.*

[numerical answer: 2016] *The question, and therefore the answer, must necessarily be qualified by the lack of a true definition of WMD. Cyber terrorism is the most likely vehicle for successful, sustained attack by a lone wolf.*

[numerical answer: 2020] *If cyber terror is included (it ought to be) then the answer is now.*

Question 4: Where do you think an actual lone wolf terrorist attack of the sort described in Question 3 might first occur?

Responses received: 46

The preponderance of opinion identifies three locations for SIMAD attacks: North America, Middle East, and Europe, with more than half of all respondents selecting North America.

Region	Responses
North America	25
South America	0
Asia (ex China)	3
China	0
Africa	1
Europe	8
Middle East	9

Confidence levels were also higher than might have been expected:

Confidence	Number of Responses
Very high	3
High	11
Middle	18
Low	6
Very low	5

Text responses:

business as usual

Due to the emotional aspect of local conflicts

The reason for my response is that post-9/11 security of large gathering venues in the U.S. has been significantly tightened (The Boston Marathon of 4/15/2013 not withstanding). The same does not appear to be the case in much of Western Europe. Almost anywhere in Europe at any point in time, have concentrations of people of mid-to-high income levels--suitable target composition to warrant high media coverage.

Americans reading about/viewing media coverage will more easily identify with the victims & will clamor for government to "do something."

Easier to hit ASIA or Europe than North America or China due to surveillance, open structure, and excellent media coverage. Harder to hit North America or China-china has lower media coverage and media suppression; US easier due to open society and cell phone and media coverage; but Russia easier due to radical and diverse ethic conflicts-Sochi will be a big target from radical Islamists. Africa will be very easy due to poor security resources, high casualties , but lower media coverage. Middle East easy targets-high risk and greater ethic, religious and nationalistic tensions.

The past is not necessarily a guideline for this question. There are strong recent historical reasons for postulating such an attack on a North American (combined US-Canada) target, given the interconnectivity of the grids, but equally there are strong reasons to believe such attacks could occur elsewhere, particularly in Russia (because of Islamist/khanate region disaffection), and also in the PRC. There is a high probability of sponsored (i.e.: not lone wolf) cyber action against, for example, Japan or the ROK.

Based on large population and geographic base, conflict is significant. Access to means could be high. Visibility can be moderate to high. Security systems to prevent may be porous. Belief systems are strong and polarizing.

I had a few criteria that drove me to select Asia. Frankly, I was guessing on meaningful criteria AND their predictive value AND how they might be relevant in Asia as opposed to the Middle East, America, Africa or Europe. In any case, a retrospective analysis might create a database that can provide such a rating. Meanwhile, I think my confidence has to be low.

Question 5: However distorted the thinking, what do you think is the single most important motivation that will fuel lone wolf attacks over the next decade?

The 45 respondents to this question favored two motivations as indicated below

Motivations	Responses
Religious incentives	19
Installing new forms of government	2
Seeking a place in history	2
Redress of perceived wrongs	17
Insanity	3
Raising money and conscripts	1
Other	1

Confidence was again quite high:

Confidence	Responses
Very high	5
High	21
Middle	15
Low	2
Very low	0

Text responses:

psychosis

I would have preferred to put two answers insanity and religion: using the Noah and the flood story as insane rationale

Even if you check religious incentives or redressing perceived wrongs, insanity is still behind this kind of aberrant action

Ideology & fanaticism coupled with perceptions, whether accurate or not, are effective self-motivators.

Seeking redress can be motivated by religious, political, or personal issues. If you disallowed that generic answer, I would chose religious.

Question 6: What do you think the primary target of lone wolf attacks will be over the next decade?

The 47 respondents to this question selected primary two targets as indicated below

Targets	Responses
Population at large	21
Specific population segments	16
Agriculture	0
Infrastructure	5
Government officials	4
Other	1

Confidence was again quite high:

Confidence	Responses
Very high	6
High	18
Middle	16
Low	3
Very low	0

Text responses:

If infrastructure means government buildings, then that would be my answer. However, if it means things like power facilities, it would not. I see the Murtha Building attack as more or less representative. An attack in Wall Street would be comparable--not against named people or civilians at large, but against "Wall Street."

I have answered population at large, but agriculture runs a close second, particularly if we imagine a binary anti-crop weapon, which can hold harvests at ransom. Terrorists hold the antidote to the poison they have administered to a crop and they extort money or demand actions to allow the harvest.

In the transforming environment in many countries, frustration at government actions/inactions has generated anger; Nigeria is a classic case. Where government targets are unavailable, public symbols and social groups will substitute.

Regarding critical infrastructure as the target, see question 3 comments.

Question 7: Do you think serious attempts to search for lone wolf terrorists who are capable of carrying out an attack using a weapon of mass destruction will be made before such an attack occurs?

Respondents: 46

Response	Responses
Yes	33
No	13

Confidence was again quite high:

Confidence	Responses
Very high	9
High	20
Middle	12
Low	3
Very low	0

Text responses:

Such means exist today and will be approved in the future.

FBI is already using trip wires (sales/distributors of hazardous chemicals and precursor chemicals) in place to provide information on suspicious sales.

Tough question to answer without context - I assume most LW terrorists will be sought in advance only after a tip has come in.

10,000 is a high number. I suggest to reduce to 500.

Question 8: What technology is likely to be most effective for the detection of people with evil intentions? Consider fields such as psychology, brain imaging, observation of unusual behavior, etc.)

Respondents: 45

Detection Techniques	Responses
Mass psychological screening	4
Monitoring of purchases of critical materials	18
Monitoring communications and social media	14
Third-party reports of unusual behavior	6
Brain physiology	0
Genetic screening	3
Other	0

Confidence was again quite high:

Confidence	Responses
Very high	10
High	13
Middle	13
Low	6
Very low	0

Text responses:

Social Signal Processing

All of the above, although in the case of a SIMAD threat, I think mass psychological screenings will be employed.

The state of Connecticut's report on the Shootings at Sandy Hook Elementary School list these observations about the shooter; can any of these, or the assemblage of all of them, form the basis for an early warning system designed to identify potential SIMADs: fascination with mass killings, playing violent video games, difficulty in social communications, possibility of being bullied as a child, proficiency with firearms, seizures, lack of emotional connections, and obsessed about battles, war, and destruction. Similar lists might be derived from studies of other mass murderers. Also consider that there may be a genetic component so that DNA analysis may play a part- or analysis of behavior of parents.

Precisely what NSA is currently accused of--eavesdropping (spying), on *everyone* is what it will take. Despite this activity running contrary to the American vision/expectation, it is the most effective means of learning what someone is thinking about, planning to do.

It is hard to give a single answer on this question. It is FAR more likely to be combinations of the kinds of techniques listed and it would be helpful if the question could reflect on those combinations.

Question 9: How successful do you think the search strategies of Question 8 will be? In a realistic future, of 100 possible lone wolf attacks how many are likely to be avoided?

Respondents: 44

Average response: 53.5% over a range that extended from 2% to 100%-- meaning from almost completely unsuccessful to completely successful.

Range of answers	Number of Responses
<20	7
20-39.9	6
40-59.9	11
60-79.9	9
>80.1	11

A graphic representation of the answers visualizes the extremely scattered views of the participants about how successful aversion strategies might be:

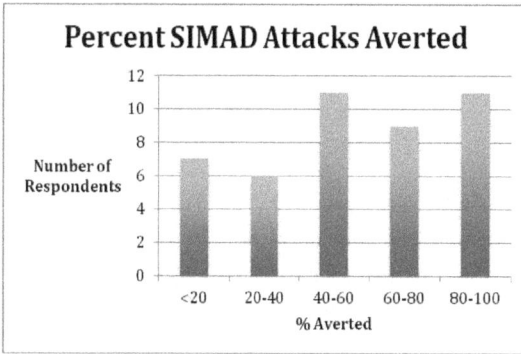

Confidence was again quite high:

Confidence	Responses
Very high	2
High	14
Middle	18
Low	7
Very low	2

Plotting the participants' confidence against the judgments about the chances for avoiding a SIMAD attack reveals a certain correlation, with a tendency for those respondents who felt there was a lower chance of averting a SIMAD attack to have lower confidence in their answers:

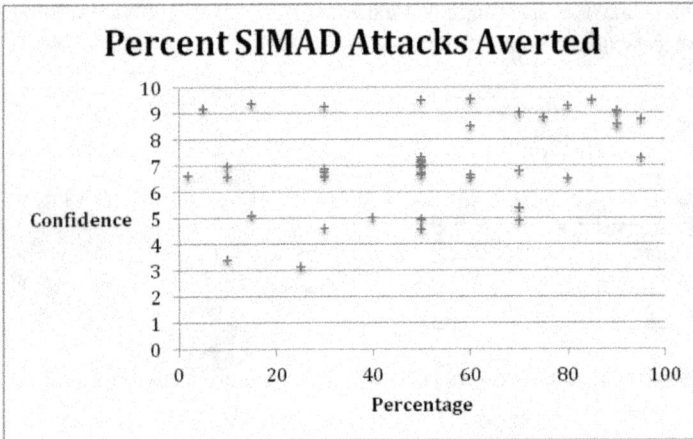

Text responses:

More at first, fewer later

Add the e-mail & social media (Face Book, Twitter, etc.), postings of the Tsarnaev brother been monitored *before* the 4/15/2013 tragedy--rather than after--the FBI would have had corroboration of what the FSB had pre-bombing, provided. This would have brought about greater in-depth investigation that might well have prevented the Boston Marathon bombing.

Question 10: Do you believe that scientific and technological papers and other publications that contain information potentially useful to terrorists should be controlled or withheld?

Respondents: 43

Responses	Number of Respondents
Yes	18
No	25

Confidence was again quite high:

Confidence	Responses
Very high	13
High	19
Middle	7
Low	2
Very low	0

Text response:

[yes] *This is a policy which could be implemented and ought to be implemented immediately*

[yes] *This had happened in Medicine for centuries*

[yes] *This way we can gain more knowledge!*

[no] *Because the very nature of scientific research is in its publication*

[no] *Societies which suppress information tend to fare badly; this is the essence of the dominating state which in itself engenders the violent response it wishes to suppress.*

Question 11: Assume that some lone wolf terrorists choose to use massively destructive or disruptive weapons; from what fields might these weapons come?

Respondents: 47

Technological Field	Responses
Nanotechnology	2
Biotech and synthetic biology	27
Nuclear physics	2
Computers/communications	6
Power generation and transmission	2
Agriculture and food	4
Other	4

Confidence was again quite high:

Confidence	Responses
Very high	6
High	21
Middle	15
Low	3
Very low	0

Text responses:

I think they might go for very simple weapons that have mass killing implications, such as very effective poisons, distributed in novel ways.

Many lone wolves are low-tech

Chemistry (and related fields), since chemical weapons of mass destruction are comparably easier to develop than biological weapons (see for example the Tokyo subway attacks)

All of the above. Intelligence is not the product of a single source, but multiple sources.

Because it is rather low-tech and relatively easily available

This question might benefit from being able to order your answer to reflect probabilities or plausibility

Biological information expanding at a very rapid pace. Not all nations follow BTWC-hence ease of distribution of bioweapons. DIY biology and synthetic biology make development of bioweapons much easier.

Invasive species as a bioweapon VERY easy to make, distribute, and have severe impact on nations, public health, ecosystems, global food production, and commodity markets.

Nanotech produces agents that are penetrative, saturational, invisible, destructive, and dispersible. Potentially, biggest bang for the buck, even more so than biological agents though nano can be combined with bio.

Question 12: Are "soft" approaches such as education reform or public awareness campaigns likely to be effective in dealing with the lone wolf threat in the long term?

Respondents: 45

Response	Number of Respondents
Yes	28
No	17

Confidence was again quite high:

Confidence	Respondents
Very high	7
High	20
Middle	13
Low	4
Very low	0

Text responses:

Not by themselves, they can reduce the problem, as technical means and public mental health can reduce the problem. Why not just say approaches like, instead of "soft" which can prejudice the unconscious.

"See something, say something" is an effective public awareness tool. "Click it or Ticket it," has reduced automobile accident injuries in every U.S. state by a significant percentage as compared to such pre-public service announcements.

Helpful if we had a context for "effective." Even one prevention is worth it, but I presume you mean effective on a larger scale.

There should be more options to answer this question, not a simple yes or no!

Long term, probabilistic influence but unless the global system undergoes a dramatic transformation, there will always be people sufficiently aggrieved or twisted to utilize available technology for purely destructive ends. But I do believe that if the available pool of such people is reduced by improving social conditions and building community, the result will be fewer violent attacks.

While soft approaches might work fairly well in most cases, inevitably they will not be 100% effective and therefore the threat will exist still

Yes - such approaches are probabilistic and long-term but will result, over time, in a smaller pool of people willing to engage in lone wolf terrorism. But unless the global system undergoes dramatic change, there will always be those aggrieved or twisted enough to pursue violence, even if only for its own sake.

Question 13: Are actions which intrude on privacy of people or otherwise compromise their civil rights justifiable in view of the threats?

Respondents: 46

Response	Number of Respondents
Yes	24
No	22

Confidence was again quite high:

Confidence	Respondents
Very high	11
High	18
Middle	6
Low	3
Very low	0

Text responses:

[yes] *Yes, because of the potential damage.*

[yes] *Yes, when so much is at stake.*

[yes] *yes, these are applied today for less important reasons*

[yes] *Yes. There will always be the need to find a balance between security & freedom of speech & the right to know.*

[yes] *Yes, because of the potential for mass casualties.*

[yes] *only with court order similar to search warrants. These could be specially designed judicial setups*

[yes] *Justifiable, but requires considerable caution, given the fact that the response tends to endanger the values which the society attempts to sustain.*

[no] *no, because I don't think they would be really effective to counter lone wolf, but would infringe basic civil liberties (already too compromised)*

[no] *No as I consider these as taboo*

[no] *Stated this way, I have to say no. But, if the question posed some balancing mechanism or meaningful/effective oversight, I might go to yes.*

[no] *No. The resolution "Right to privacy in the digital age" adopted by the UN General Assembly in December 2013 is rightly calling on all countries to take measures to end activities such as electronic surveillance, interception of digital communications and collection of personal data, which violate the fundamental "tenet of a democratic society."*

[no] *No, who will control the controllers?*

Question 14: If people are identified as potential lone wolves of the sort we have already seen currently, how do you think society should deal with them?

This was the first of four open ended questions that allowed respondents to answer in any way they chose. In all, more than 52 suggestions were received. With few exceptions it was possible to group the responses into the following major categories:

Category	Percent
Incarceration, removal from society	35
Rehabilitation, counseling, and psychotherapy	17
Monitoring, observation, placing on a watch list	33
Medical treatment	2
Due process	8
Other	5

The verbatim comments and suggestions were:

Monitoring and if justified, then incarceration and psychotherapy.

Incarceration

Information collection and in end incarceration.

If they are indeed identified, they should be dealt with

Incarceration and, if necessary, therapy.

Yes; should be jailed.

Incarceration. [several answers]

"Potential" is a potent word in this question. We cannot lock people up for what they might be thinking absent proof of intent to commit. Monitoring is a balanced approach. This is a hard one to answer, because most of us would prefer (I think) to give contextual answers.

If potential lone wolves are identified, they should be used to learn more about their psychology. They should be monitored and interrogated about their beliefs, societal approaches and miscontentments, and asked about their suggestions for solutions to address the problems. Then further interrogation on their eventual motivation, scope and spectrum, and potential action(s) considered. If justified, then isolation and potential treatment should be considered.

There need to be efforts made at radicalization prevention and improvement of acculturation dynamics.

Monitor but also work to have them get mental health

Monitoring and case-dependent action when justified

They should be monitored and incarcerated

just as any psychotic individual they should be removed from society for their own good and the good of society....but they retain their rights as individuals

observed, monitored, interrogated and maybe isolated in the very end only

Monitoring followed by incarceration based on the threat level.

They should be monitored and medically treated.

Community support

Psychological counseling

Assisting in re-channeling i.e. finding something meaningful but benevolent to

The problem is "identified": by whom and on what basis? I think the basis should be solid investigative work - and without the shabby entrapment strategies often used. If someone can be shown to have entered upon a path of accumulating real data and materials for such an attack, and has entered into communication regarding such an attack, then they should be arrested and prosecuted for the actual laws they have broken (i.e., as a criminal not a terrorist). This is the best way to protect society not only from attacks but from its own authoritarian tendencies.

Information collection and possible incarceration, but only if conspiracy can be proven. We have seen that incarceration often actually creates viable threats, and institutions in the US and Europe have tended to be the major recruiting grounds to turn lone wolves into components of wolf packs.

If they have not broken any laws, observation. If they have or are about to commit an attack, interception and due legal process.

We should wit until "Potential" matures

Interrogations, arrest

Law enforcement should place them on watch list, not active investigation.

"Potential" is a potent word in this question. We cannot lock people up for what they might be thinking absent proof of intent to commit. Monitoring is a balanced approach. This is a hard one to answer, because most of us would prefer (I think) to give contextual answers. If they have not broken any laws, observation. If they have or are about to commit an attack, interception and due legal process.

Informing deans at universities, police chiefs and security employees about the risks.

monitoring

Education and social rehabilitation

If someone decides to act as a lone wolf, there is little opportunity for the law enforcement to change his/her mind. The most useful measure would be, in this situation, monitoring and, if justified, isolating him/ her.

Monitoring, evaluation, isolation is necessary.

Question 15: What if they have been engaged in building weapons that could produce mass destruction or disruption?

As in the previous question respondents were free to answer in anyway they chose. The categories of suggestions were essentially the same as before; however in this instance attitudes seemed to be tougher.

The verbatim comments and suggestions were:

Interrogation and discussions with relevant scientists and security experts to learn more about lone wolf tactics; monitoring and eventual incarceration and treatment if possible.

Monitoring and control

Information collection

Awareness, education, monitoring, and control

monitoring, regulations, cooperation with science community, social media, develop tech means to detect evil intensions

Not enough or not soon enough, will always haunt us. Totalitarian societies seldom have this dilemma.

Incarceration for life.

Since there are few benign uses of such weapons, then they can be arrested for conspiracy to commit.

Interrogated about their motivation, scope and spectrum of the potential attack, potential network of which they might be part; source of their know-how for weapons construction, etc.; plus monitoring and eventual isolation and treatment.

increased government monitoring in conjunction and compliance with domestic and international law

closely monitored and be prevented from returning to making new mass weapons of destruction.

monitoring and, if justified, case-specific action such as incarceration and/or therapy

incarceration for life

the same

consider the possibility of special courts/tribunals in which the judges include psychiatrists to examine and prescribe for the suspect

monitoring, close observation, interrogation and then maybe isolation

Monitoring followed by incarceration and even elimination, as warranted.

Community support Psychological counseling Assisting in re-channeling i.e. finding something meaningful but benevolent to do Limiting ability to produce weapons of mass destruction at the global level, i.e. those should not exist in the first place anywhere

incarceration and, if necessary, therapy.

Monitoring and medical therapy

As in previous answer, arrest and interrogation and prosecution according to the law. The red herring of "what if they were about to explode a nuclear bomb...etc." has not occurred and is a very specific, anecdotal, and unique "wild card" case that cannot be the justification for effective prevention. It only leads to torture, false data, generation of new hostiles, and degradation of society.

Monitoring and therapy

arrest interrogation

High possibilities, but also could be friends or criminals that know about this expertise

deal with them according to legislation

Legal action including incarceration

The authorities should monitor and take immediate action to prevent them from building such weapons. Education and raising public awareness can contribute also.

Action meaning isolation, interrogation, and incarceration if necessary.

Question 16: What steps, if any, do you think should be taken to minimize lone wolf threats? When?

This open-ended question led to many different suggestions. Roughly, the comments could be divided into two major camps:

- those who promote corrected social ills responsible for lone wolf behavior
- those who recommended higher level safeguards, some draconian, to minimize the threat.

About half of the 30 suggestions fell into each category.

The verbatim suggestions were:

More emphasis on ethics and compassion in education in general, and concerning potential S&T implications, specifically. When new-technology weapons are developed, the obligation to also develop the anti-dote for them or if not possible, moratorium on the use of the respective technology.

Monitoring and control

Awareness, education, monitoring, and control

also ban the use of WMDs and other future technologies in the bio-nano field

Making use of all the technologies listed in question 11 to pinpoint identifiable behavioral traits will mitigate & ameliorate but will not eliminate the threat.

I think first generation disaffected youth in the West could have interventions be successful before they turn violent. In other parts of the world, economic development and educational opportunities would be a long-term viable solution.

Intelligence gathering, both electronic and humint.

Monitoring purchases of critical materials;

Ban on weaponizable bio/nano technologies. Ban possession of weapons by citizens. Continuous surveillance of social satisfaction--who and why people are not happy and how could that be fixed.

radicalization prevention programs for youth (i.e., especially first generation)and acculturation facilitation programs would be desirable as well as continuous supervision of job acquisition and retention efforts.

Reduce situations that lead to development of lone wolves, monitor of society, reduce bullying in schools, education, etc.

Support mental health issues from childhood to ageing.....

education, societal awareness programs, mental health programs, and most importantly study of societal injustices, poverty, corruption, deprivation, abuse that may provoke radicalization

Education, public awareness, anti-lone wolf drills, provision of antidotes, monitoring / surveillance, incarceration and elimination, as necessary.

Lone wolfs are a product of a society. Social integration instead of alienation should be a goal. Compassion, mental health support, support for young males who are struggling, etc.

reduce access to potential weapons/dangerous materials (e.g. assault rifles, some medical/scientific equipment)

In the short term, monitor Internet activities, especially social networks. In the long term, promote social integration of potential lone wolves.

Long term: education and reduction of poverty; higher level public discourse. These are long-term, however, and probabilistic. More immediately: sensors at key transit points and supply sources; focused monitoring of movement of potentially destructive materials - much more effective than data sweeps and tracking every blip in cyber-space. Good police work that responds to actual forays into purchase of material or explicit threats on social networks, etc.

The question perhaps should be turned around: why do people get angry? Is this an inevitable function of urbanization, economic dislocation and perceived removal of opportunity for various groups, the inevitable result of a sclerosis in modern social structures over time (i.e.: the maturing of a civilization with a detritus of legislation and governance)? The answer in some parts lies in community response (i.e.: reaction) to an immediate security threat; and in some parts lies in transforming the society to minimize the perception of overbearing government where that exists.

A greater level of public awareness and education. Addressing substantially those social, political and economic factors that drive the problem. This should be in conjunction with monitoring.

Monitoring and control

Intelligence, surveillance,

Trip wires, to watch access to WMD and precursor material, and solicit the appropriate communities to be aware of threats emerging from those communities.

Awareness, education, monitoring, and control

1. Check family environment. 2. Check main problems inside communities 3. Check governments and security; 4. Check TOC people.

monitoring, awareness raising, education

Improve social and health care, increase labor opportunities, and reduce inequality.

Social education, 3rd party monitoring

Minimizing this threat is a long time continuous effort for the national and international authorities that imply economic, social, moral, political, educational factors. Soft and hard measures should be combined according to the specific situations. There is no magic formula to stop this phenomenon.

First - education, to prevent. Second - monitoring to know Third - isolation and incarceration, if the danger is high.

Question 17: What steps, if any, do you think should be taken to minimize threats that may achieve destruction on the scale depicted in Question 2? When?

In this final question we were interested in receiving suggestions from the panel about actions that might be taken if SIMAD catastrophes where to become even more plausible and tangible; in other words, suggestions for heroic efforts at deterrence.

The verbatim suggestions were:

If the threat is severe enough, maybe after the first SIMAD catastrophe has occurred, a new version of brain washing might be considered to physically change the brain of potential SIMADs. Can it be done? fMRI can give us pictures of the brain and in the future perhaps highlight abnormalities.

Additionally it is possible that a genetic component exists which induces aberrant behavior. If that is so, genetic manipulation might also be considered. Of course this brings important civil liberty questions to the surface.

Set up a system where marginally crazy people, under close control and observation, are asked to generate ideas for weapons that a SIMAD might use

Do you mean "when" in terms of society-wide approaches? Early education, materials monitoring, key word searches outside of US.

This question seems to imply that scale of threat might permit a broader scale of steps to minimize the threat - is that the intent?

More emphasis on social justice and ethics at all level. The media should not "hero-ise" those that commit mass destruction or aggression of any kind. Monitor social media for up-normal behavior to identify potential lone wolves. Enact and enforce international regulations for punishing crimes and the use of WMD.

Increased social media monitoring, identify patterns, have special forces trained in this domain.

Due diligence with respect to weapons supply flows; radicalization prevention and acculturation facilitation efforts; coordination of government and non-state actor (NGO, CBO) integrative efforts.

Put security forces on alert, have additional training for mass causalities, review training for civilian and military communication and cooperation during WMD events. Assess lone wolves and their potential and their Internet usage and purchases. Increase HUMINT on weapons development and interactions with lone wolves.

I see no change ...nor do I see that we can stop all untoward occurrences in society...the price to freedoms would be too high ....monitor all the obvious places that you name .. and support mental health programs globally.....and support unstable societies long term ..with nonviolence.... for generations and allow them to develop

Same as answer to Question 16.

Improved information sharing between intelligence, police and armed forces. Greater access to detection tools for vulnerable target.

As in the previous question, Web monitoring and control in the short term and social integration in the long term.

For high-level threats, we need flexible and responsive investigative capabilities that are not sidetracked by bureaucracy, as was the case with the Boston Marathon bomber. All the long-term solutions apply but for an imminent threat, we need network-based tactics that can prevent a target from doing maximum damage even when the perpetrator is as yet unidentified.

SIMAD major actions should trigger a concern that all is not right in a society, or in its relations with another society. But just because a lone wolf action may inflict major damage does not necessarily mean that the action reflects the will of views of an entire section of society. The actor may be aberrant. Did the attack on Archduke Ferdinand in Sarajevo warrant the start of World War I, for example? Minimizing media hysteria is the key to removing the appeal of all forms of terrorist behavior, including lone wolf actions, but can media hysteria be reduced without impinging on fundamental freedoms? Or can media responsibility be developed? Or is it merely normal crowd behavior that we see reactive hysteria in the face of attacks which are called "terrorism" because they engender a reaction called "terror"?

More emphasis on social justice and ethics at all level. ASAP

aggressive

Same answer as question 16. trip wires, watching materials that can be used to construct WMDs and recruiting members of communities to be self aware of potential threats emerging from those communities.

Put security forces on alert, have additional training for mass causalities, review training for civilian and military communication and cooperation during WMD events. Assess lone wolves and their potential and their Internet usage and purchases. Increase HUMINT on weapons development and interactions with lone wolves.

1. Teach values. 2 Show people that human rights are respected. 3. Media should be full time involved in sending subliminal messages, so people can reflect about their own behaviors.

Raising security measures, closely monitoring the (potential) terrorists, understanding the drivers and patterns that can influence such behaviors.

Is necessary to understand the nature of the reasons of SIMAD, to connect to the changes of the social environment and to control the possible tool, technologies or substances that can be used in terrorist attacks.

Other Comments Made by the Respondents

Comments on the design and the process:

Great questionnaire!

Congratulations. The questionnaire and the system is working very well. Finally Infinitum server in Mexico solved today the access problem to this questionnaire.

Interesting subject.

I think that also multiple choice answers should be made possible.

Questions 14-17 should include notification of being answered (as existing in the previous questions)

Impossible to answer the last 3 questions. The link did not work for me.

Comments on substance:

Do you think that this threat links to social integration of minorities? What are the early signs of this threat? How can we monitor possible threats? Do you think this threat needs different and new regulations?

Research into invasive species as next generation bioweapons as well as class C (CDC list) bioweapons and contingencies if these weapons are used must be researched and funded.

Who is monitoring all the "caretakers" .at present and after they retire......who have access to nuclear, medical research, nano etc. materials? The H5N1 controversy is a good example... and I am concerned about "private" science as a whole....who monitors Craig Venter?

Study radicalization phenomenon in general. Societal injustices, perceived threats, deprivations, hurt, actual loss, alienation, sense of loss, sense of pain. Perceptions matter and reality is perceived differently by various people. Group identification of loss of dignity, pride, status, or destiny is important. The emphasis is on group study to get to individual behavior. After all, no lone wolf acts in this blue air. More research effort and dialogue is needed to comprehend and then tackle the issues

More Questions on Cyber crime. We will find the wolfs there....

I think more emphasis should be put on lone wolf cyber-terrorists

Appendix C. Demographics of the Participants in the RTD

The following graphs indicate the demographics of the participants in the Real-Time Delphi.

Regional demographics:

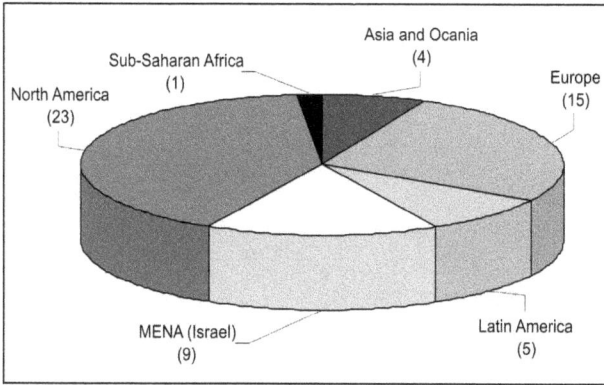

Sectoral demographics:

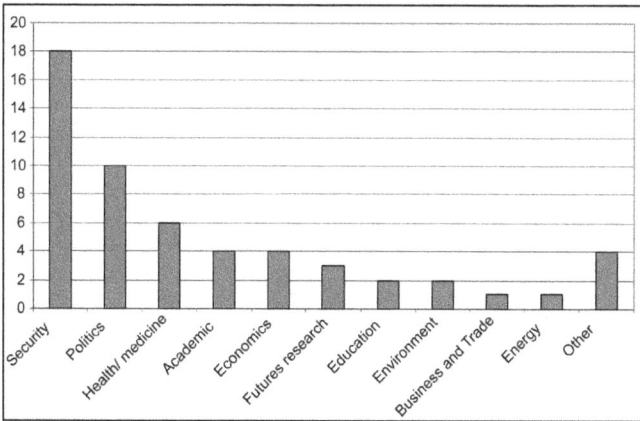

Gender distribution
    Males: 83%
    Females: 17%

The list of participants with their location and institutions is available on the study's webpage: http://www.millennium-project.org/millennium/SIMAD.html. We are very grateful for their valuable inputs. Note: none of the conclusions of the RTD study or statements in this book should be attributed to any of the participants.

Appendix D. The Real-Time Delphi Method

The Delphi method was originally developed at the RAND Corporation in the late 1950s and '60s as an effective means for collecting and synthesizing expert judgments. The technique requires sequential questionnaires, with information feedback from round to round. Although it has been very widely used since the first Delphi study was conducted and published by RAND in 1964, given the several rounds of questionnaires, to complete a whole study might take a long time. Nevertheless it is a principal method of futures research and has found application in planning, decision- making, and policy research.

In September 2004, the Defense Advanced Research Projects Agency (DARPA) provided a grant to Articulate Software, Inc. to develop a Delphi-based method for improving the speed and efficiency of making tactical decisions under conditions of high uncertainty, using expert judgments in a Delphi-like process. The first Real-Time Delphi design (developed by T. Gordon and A. Pease) came out of this work. The new process is "roundless" and can-- if desired--be accomplished in a much shorter short time than the earlier forms of Delphi. Since it is online, communication times are short and participation can be asynchronous.

The technique has been used to good advantage many times since its introduction. The Millennium Project developed the code from the initial open source versions and has used it in numerous studies such as:
- Elements of Future Financial System
- Elements of an Energy Future
- Arts and Media Futures
- Education 2030
- Latin American Scenarios
- Education Requirements for Ontologists
- Changes to Gender Stereotypes
- Azerbaijan: Key Future Developments and Measures
- State of the Future Index (for several countries)
- The Future of Cooperatives and other Business Forms
- Argentinian Agricultural Policy
- Selecting topics for inclusion in a UNESCO global water report.
- Assessment of Potential Strategies (Global Environmental Forum)
- Voting for MP Board Members

How does the RTD Work

Imagine a Real Time Delphi study involving a set of questions. When a respondent joins the ongoing study, he or she is presented an on-screen form which contains, for each question:

- The question and a place for the respondent to supply their response.
- The average response of the group and the number of responses received so far
- A window showing reasons that others have given for their responses and a place for the respondent to type in justifications for their own answer

Respondents are asked to consider all of this information and to provide their input. The computer saves the answer and the group average is updated immediately and presented back to the respondent and anyone else who has signed on.

The administrator has many design options; for example, he or she may specify that if the respondent's answer to any question is beyond a pre-specified distance from the average or the median, the respondent receivers an "attention" message (however, this does not prevent the respondent from maintaining his or her original answer.)

There are no explicit rounds as in the original Delphi process. As long as the questionnaire is open for inputs, respondents can come back to the questionnaire at any time. They are presented with their own answers as well as the updated averages and other contributions. Thus, they can see how the group's responses evolve and even edit/change their own answers.

RTD yields the distribution of the group's responses and reasons for the extreme positions. The process can be synchronous or asynchronous, and since it is usually on an Internet web site, can involve a worldwide panel. The Delphi requirements of anonymity and feedback are also respected in this process.

Appendix E. Examples of LW Cases

Databases of Terrorist and Lone Wolf Events[287]

One of the most extensive database of terrorist events is the RAND Database of Worldwide Terrorism Incidents; it can be accessed at http://smapp.rand.org/rwtid/search_form.php

This database permits filtering the recorded set by time of occurrence, country, perpetrator, tactic, weapons, and several other parameters. Each incident is accorded a paragraph or so of description. There are 40,000 entries covering events from 1968 to 2009; updating is reportedly underway. The database includes activities by both groups and individuals but there is no filter to sort the LWs from the larger set.

The Global Terrorism Database (GTD) is even larger and is maintained and made available by the National Consortium for the Study of Terrorism and Responses to Terrorism (START) at the University of Maryland, USA. It includes information on terrorist events around the world from 1970 through 2013 (to be continuously updated), comprising more than 125,000 cases. For each incident, it provides information on the date and location of the incident, the weapons used and nature of the target, the number of casualties, and--when identifiable--the group or individual responsible. It is accessible at http://www.start.umd.edu/gtd/

The New America Foundation's Database of homegrown extremism (http://securitydata.newamerica.net/extremists/analysis) lists 122 incidents that include both LW and group attacks--domestic and foreign--that have been performed by American citizens and permanent residents that have occurred or have been foiled. It contains data from 2001 to 2014 (latest update in October, 2014) and focuses on and is organized by method of detection: e.g. tip, informant, suspicious activity report, etc. This is an excellent but limited set since it deals only with U.S. citizens and residents. While it is difficult to class the incident descriptions as being performed by a LW or not, a good approximation is that about 30% to 50% of the entries fit the definition of LW.

The Heritage Foundation compiles a list of foiled or failed terrorist attacks that can be accessed at http://www.heritage.org/research/reports/2013/07/60-terrorist-plots-since-911-continued-lessons-in-domestic-counterterrorism. This listing includes 60 foiled events, mostly in the U.S.

---

[287] This list first appeared in the article "Prospects for Lone Wolf and SIMAD terrorism" published in *Technological Forecasting & Social Change* TFS-18143; © Elsevier Inc., 2015. The authors of the article and this book are the same: Theodore Gordon, Yair Sharan, and Elizabeth Florescu.

that have occurred between 9/11 and 2013. A review of this list shows that of the 60 attempts, some 22 (37%) were by LWs.

The Wikipedia entry on lone wolf attacks lists 36 LW attacks conducted between 1998 and 2014. The definitions used in this compilation are somewhat different; for example the list includes Timothy McVeigh who did not act totally alone. To the set of 36 examples provided one might add school shootings, the Sandy Hook shootings, the second Fort Hood Shooting, the New York hatchet attack of October 13, 2014, the 2014 War Memorial killing in Canada, and several others.

Petter Nesser also complied a list of 15 LW completed and failed terror attempts for further analysis, available at: http://www.terrorismanalysts.com/pt/index.php/pot/article/view/231/html. This list includes assassination attempts, plots to blow up aircraft, arson, and mailing of toxic letters.

The report, "Lone-Wolf Terrorism, A Case Study By The European Research Consortium Transnational Terrorism, Security And The Rule Of Law (TTSRL); http://www.transnationalterrorism.eu/tekst/publications/Lone-Wolf%20Terrorism.pdf contains a number of important tables:
Chronology of lone-wolf terrorism in the United States, 1940-2007 (that begins, interestingly enough with a report of a series of bombings that tool place between 1940 and 1956 in New York city)
Chronology of likely cases of lone-wolf terrorism in TTSRL sample countries, 1968-2007
Chronology of likely cases of lone-wolf terrorism in Canada and Australia, 1968-2007
Incidents of terrorism in TTSRL sample countries, 1 January 1968 – 1 May 2007 (lists 5646 incidents by country)
Hewitt, C. (2005) *Political Violence and Terrorism in Modern America* (Westport and London: Praeger Security International) presents a chronology of terrorist events from 1956 to 2005, including "date of the incident, type of the incident, the group or person responsible, where the attack occurred, and details of the act."[288]

Harvey W. Kushner's (2003) *Encyclopedia of Terrorism* is an extraordinary 532 page compendium of terror incidents, locations, methods, and perpetrators with data to October 2002.

The table that follows in this appendix uses descriptions from one or more of these sources and from blogs, news reports, TV, FBI descriptions, indictments, and police reports. In all, 100 events were selected for inclusion here; a few would fall outside of the strict definition of a Lone Wolf (e.g. Timothy McVeigh who apparently had some outside assistance). These accounted for 500 deaths, 2,029 injuries; among these 18 included the term "weapons of mass destruction" in police reports or indictments or seemed to fit in the category, and 46 were foiled.

---

[288] Christopher Hewitt, *Political Violence and Terrorism in Modern America: A Chronology* (Praeger Security International), September 30, 2005, ISBN-13: 978-0313334184

Date	W M D	Killed	Wou nded	Fail ed	Country	Perpetrator or Suspect	Circumstances	Weapons	Reference
9/16/13		13	8		US	Aaron Alexis	Mass shooting in the Washington DC Naval Yard. Shooter, a contractor, apparently had mental issues. Shot and killed by law enforcement officers.	Fire arms	http://www.defense.gov/pubs/DoD-Internal-Review-of-the-WNY-Shooting-20-Nov-2013.pdf and http://www.fbi.gov/washingtondc/press-releases/2013/law-enforcement-shares-findings-of-the-investigation-into-the-washington-navy-yard-shootings
6/1/09		1	1		US	Abdulhakim Mujahid Muhammad	Known as Little Rock recruiting office shootings, Muhammad was charged with murder and other crimes. Pled guilty and was sentenced to life in prison.	Fire arms	http://en.wikipedia.org/wiki/2009_Little_Rock_recruiting_office_shooting and http://www.investigativeproject.org/3061/little-rock-jihadist-sentenced-to-life#
2/17/11	Y			Y	US	Amine El Khalifi	Arrested on charges of plotting to attack the U.S. Capitol. Pleaded guilty to "attempting to use a **weapon of mass destruction.** Sentenced to 30 years in prison.	Fire arms, IED	http://www.fbi.gov/washingtondc/press-releases/2012/virginia-man-sentenced-to-30-years-in-prison-for-plot-to-carry-out-suicide-bomb-attack-on-u.s.-capitol and http://www.nydailynews.com/news/national/amine-el-khalifi-sentenced-30-years-capitol-bomb-plot-article-1.1159847
6/5/03	Y				Belgium	45 year old Iraqi man	Mailed letters containing a nerve gas ingredient; to Belgian prime minister's office, and the U.S., British and Saudi Arabian embassies	Toxins in letters	http://www.freerepublic.com/focus/news/923015/posts and http://www.democraticunderground.com/discuss/duboard.php?az=view_all&address=102x205890
6/22/11	Y			Y	US	Abu Khalid Abdul-Latif	Thought to have been planning to kill people in the Military Entrance Processing Station as revenge for U.S.in Afghanistan, prosecutors said. Pleaded guilty to conspiracy to murder U.S. officers and use a **weapon of mass destruction.** Sentenced to 18 years in prison.	Machine gun, grenades	http://www.huffingtonpost.com/2013/03/25/abu-khalid-abdul-latif-se_n_2950909.html and http://www.fbi.gov/seattle/press-releases/2013/seattle-man-sentenced-to-18-years-in-prison-for-plot-to-attack-seattle-military-processing-center

Date	WMD	Killed	Wounded	Failed	Country	Perpetrator or Suspect	Circumstances	Weapons	Reference
12/14/12		28	2		US	Adam Lanza	Sandy Hook shooter in school shootings in Newtown CT. Committed suicide when first responders arrived.	Fire arms	http://en.wikipedia.org/wiki/Sandy_Hook_Elementary_School_shooting and http://www.ct.gov/csao/lib/csao/Sandy_Hook_Final_Report.pdf
9/14/12	Y			Y	US	Adel Daoud	Arrested for attempting to bomb a bar in Chicago; charged with attempting to use a **weapon of mass destruction.** FBI sting. Sentence 23 years in prison.	Bomb	http://www.investigativeproject.org/case/624
5/12/11				Y	US	Ahmed Ferhani	Plotted to attack New York synagogues. Pleaded guilty to terrorism charges; Received a 10 sentence.	Bomb	http://www.huffingtonpost.com/2013/03/15/ahmed-ferhani-man-who-plead-guilty-nyc-synagogue-bomb-plot_n_2883836.html and http://www.nytimes.com/2012/12/05/nyregion/ahmed-ferhani-pleads-guilty-in-plot-to-blow-up-synagogue.html?_r=0
3/6/08		9	11		Israel	Alaa Abu Dhein	Attacked people in a Jewish seminary in Jerusalem. Dhein shot dead at scene.	Fire arms	http://www.iris.org.il/blog/archives/2720-Released-Terrorist-Kills-8-at-Jerusalem-Yeshiva.html and http://www.tlvfaces.com/terror-victim-attacked-terrorists-way-sons-grave/
2/23/97		1	6		US	Ali Hassan Abu Kamal	Shooter at observation deck of the Empire State Building. Committed suicide after shootings	Fire arms	http://www.cnn.com/US/9702/24/empire.shooting/ and http://murderpedia.org/male.A/a/abu-kamal-ali-hassan.htm
7/22/11		77	319		Norway	Anders Behring Breivik	Two consecutive attacks. 1) a car bomb in front of Norwegian government headquarters in Oslo. 2. At the island of Utøya, where he shot for 90 minutes. Surrendered to police and is now in prison for a term that may be life.	Bomb, Fire arms	http://en.wikipedia.org/wiki/Anders_Behring_Breivik and http://www.nytimes.com/2012/08/25/world/europe/anders-behring-breivik-murder-trial.html
7/22/11		1			US	Andrew Hampton Mickel	Convicted of murder. Postings on Internet read "Hello Everyone, my name's Andy. I killed a Police Officer in Red Bluff, California...." and signed with an alias of Mickel's. Sentenced to death.	Fire arms	http://murderpedia.org/male.M/m/mickel-andrew-hampton.htm and https://www.newsreview.com/chico/mystery-of-andrew-mccrae/content?oid=12067

Date	WMD	Killed	Wounded	Failed	Country	Perpetrator or Suspect	Circumstances	Weapons	Reference
4/17/08				Y	UK	Andrew Ibraham	Found guilty of planning to bomb a shopping center. Sentence: life in prison with a minimum of ten years.	Suicide vest Bombs	http://www.telegraph.co.uk/news/ukne ws/law-and-order/5850684/Andrew-Ibrahim-a-Muslim-convert-guilty-of-planning-terror-attack.html and http://www.theguardian.com/uk/2009/jul /17/isa-ibrahim-bristol-bomb-guilty
2/18/10		2	13		US	Andrew Stack III	Anti-Internal Revenue Service; published a manifesto before the suicide attack with a light plane. Killed in the crash.	Aircraft	http://en.wikipedia.org/wiki/2010_Austin _suicide_attack and http://www.nytimes.com/2010/02/19/us/ 19crash.html
12/10/10	Y			Y	US	Antonio Martinez	Pleaded guilty to planning to bomb a Maryland recruiting center and attempted use of a **weapon of mass destruction**. Undercover sting. Sentenced to 25 years in prison.	Bomb	http://www.fbi.gov/baltimore/press-releases/2010/ba120810.htm and http://www.investigativeproject.org/cas e/505
9/25/14		2			US	Anton Nolen	Nolen accused of attacking two women allegedly injuring one and beheading the other; citing Muslin injunction against women. Nolen shot at the scene but survived.	Knife	http://www.foxnews.com/us/2014/09/30 /oklahoma-beheading-suspect-likely-radicalized-behind-bars-say-experts/ and http://www.huffingtonpost.com/2014/09 /26/alton-nolan-beheads-cowor_n_5888500.html
3/4/11		2	2		Germany	Arid Uka	Serving a life term for killing two US soldiers on their way to Afghanistan	Fire arms	http://www.dailymail.co.uk/news/article-1362715/Arid-Uka-Frankfurt-airport-shooting-suspect-admits-targeting-American-troops.html and http://www.bbc.com/news/world-europe-16984066
8/17/05		4	2		Israel	Asher Weisgan	An Israeli bus-driver, protested Israel's Gaza disengagement plan; convicted of killing four Palestinians	Fire arms	http://murderpedia.org/male.W/w/weisg an-asher.htm and http://www.ynetnews.com/articles/0,73 40,L-3308882,00.html
2/2/96		3			US	Barry Loukaitis	Convicted of killing three fellow students in school shooting in Moses Lake WA. Thought to be inspired by Pearl Jam video. Serving two life sentences without parole.	Fire arms	http://www.karisable.com/loukaitis.htm and https://schoolshooters.info/sites/default/ files/Loukaitis%20court%20document.p df

Date	W M D	Killed	Wou nded	Fail ed	Country	Perpetrator or Suspect	Circumstances	Weapons	Reference
2/25/94		29	100		Israel	Baruch Goldstein	American born Israeli physician killed Muslims in prayer in the Cave of the Patriarchs. Beaten to death at the scene.	AK 47	http://murderpedia.org/male.G/g/goldst ein-baruch.htm and http://en.wikipedia.org/wiki/Cave_of_th e_Patriarchs_massacre
7/2/99		2	9		US	Benjamin Nathaniel Smith	A white supremacist who went on a killing spree targeting minorities. Committed suicide.	Fire arms	http://murderpedia.org/male.S/s/smith- benjamin-nathaniel.htm and http://www.wsws.org/en/articles/1999/0 7/kill-j06.html
6/23/05		5	doze ns		US	Bruce Ivins	Suspected mailer of anthrax letters; investigating committee said it could not reach a "definitive conclusion of the origins of the (anthrax) in the mailings based on the scientific evidence alone."	Anthrax letters	http://en.wikipedia.org/wiki/Bruce_Edw ards_Ivins and http://topics.nytimes.com/top/reference/ timestopics/people/i/bruce_e_ivins/inde x.html
8/10/99		1	5		US	Buford O. Furrow, Jr.	Pleaded guilty to shootings in an attack on a Jewish daycare in Los Angeles; Furrow was characterized as a member of a white supremacist group. Sentenced to two life terms in prison plus 110 years	Fire arms	http://www.cbsnews.com/news/furrow- many-signs-of-trouble/ and http://murderpedia.org/male.F/f/furrow- buford.htm
1/14/15				Y	US	Christopher Cornell	Indicted for attempted murder, solicitation to commit a crime of violence and possession of a firearm. Pleaded not guilty. No hearing yet. FBI sting.	Pipe bombs Fire arms	http://www.csmonitor.com/USA/Justice/ 2015/0115/How-alleged-lone-wolf- terrorist-plotted-attack-on-US-Capitol- and-was-stopped and http://www.foxnews.com/us/2015/03/07 /ohio-terrorism-suspect-wanted-to- shoot-obama-in-head/
2/4/95		3	116		England	David Copeland	Three bombings. Known in press as "London Nail Bomber"; held neo-nazi views and that he wanted to start a race war. Convicted of three murders and planting bombs. Sentenced to 6 life sentences.	Nail bombs	http://murderpedia.org/male.C/c/copela nd-david.htm and http://www.theguardian.com/uk/2000/ju n/30/uksecurity.sarahhall
12/6/06	Y			Y	US	Derrick Shareef	Charged with attempted use of a **weapon of mass destruction** and sentenced to 35 years in a plot to set off hand grenades in an Illinois shopping mall.	Hand grenades	http://www.justice.gov/archive/opa/pr/2 008/September/08-nsd-872.html and http://abcnews.go.com/TheLaw/story?i d=2710776

Date	W M D	Killed	Wou nded	Fail ed	Country	Perpetrator or Suspect	Circumstances	Weapons	Reference
6/7/03		3			US	Devin Moore	Sentenced to death for killing 2 policemen and a dispatcher; thought to have been influenced by the game Grand Theft Auto	Fire arms	http://murderpedia.org/male.M/m/moor e-devin.htm and http://www.tuscaloosanews.com/article/ 20120218/news/120219745
4/15/13	Y	3	250		US	Dzhokhar and Tamerlan Tsarnaev	Boston Marathon killings. Dzhokhar Tsamaev defense admitted to his planting bombs. His brother was killedafter the bombing	IED bombs	http://www.history.com/topics/boston- marathon-bombings and http://en.wikipedia.org/wiki/Tamerlan_T sarnaev
8/4/05		4	12		Israel	Eden Natan-Zada	Israeli soldier, a deserted after refusing to move settlers from West Bank; opened fire in a bus. Killed by a mob after shooting,	Fire arms	http://www.washingtonpost.com/wp- dyn/content/article/2005/08/04/AR2005 080401350.html and http://www.jewishpress.com/tag/eden- natan-zada-opened-fire-on-a-bus/
5/20/13					US	Edward Snowden	Charged with violating the Espionage Act and theft of government property.	IT	http://en.wikipedia.org/wiki/Edward_Sn owden and http://www.theguardian.com/world/201 3/jun/17/edward-snowden-nsa-files- whistleblower
1/4/11				Y	US	Emerson Winfield Begolly	According to the FBI, "Begolly systematically solicited jihadists to use firearms, explosives, and propane tanks against targets such as police stations, post offices, Jewish schools and daycare centers, military facilities, train lines, bridges, cell phone towers, and water plants." Sentenced to 8 1/2 years in prison.	Fire arms, IT	http://www.fbi.gov/pittsburgh/press- releases/2013/pennsylvania-man- sentenced-for-terrorism-solicitation- and-firearms-offense and http://www.huffingtonpost.com/2013/07 /16/emerson-begolly-found- guilty_n_3605876.html
1996-1998		3	150		US	Eric Rudolph	Pleaded guilty to homicide; multiple bombings. Targets included abortion clinics, gay clubs, and the 1996 Olympics in Atlanta. Sentenced to four life terms plus 120 years in prison.	Bombs	http://www.crimelibrary.com/terrorists_s pies/terrorists/eric_rudolph/6.html and http://www.fbi.gov/news/news_blog/this -day-in-history-eric-rudolph-captured

Date	WMD	Killed	Wounded	Failed	Country	Perpetrator or Suspect	Circumstances	Weapons	Reference
5/1/10	Y			Y	US	Faisal Shahzad	Times Square bomber. Pled guilty to all charges against him including conspiracy to commit an act of terrorism and to use a **weapon of mass destruction.** Sentenced to life in prison without parole.	Car bomb	http://www.justice.gov/opa/pr/faisal-shahzad-pleads-guilty-manhattan-federal-court-10-federal-crimes-arising-attempted-car and http://www.nytimes.com/2010/10/06/nyregion/06shahzad.html
10/27/10				Y	US	Farooque Ahmed	Arrested for allegedly plotting to attack the Washington, D.C. subway. Pled guilty to terrorism charges. Sentenced to 23 years in prison. FBI sting.	Bombs	http://af.reuters.com/article/worldNews/idAFTRE73A6JE20110411?sp=true and http://www.fbi.gov/washingtondc/press-releases/2010/wfo102710.htm
10/29/94				Y	US	Francisco Martin Duran	Convicted of attempted assassination of President Bill Clinton; fired 29 shot from the fence surrounding the WH. Sentenced to 40 years of prison time.	Fire arms	http://tech.mit.edu/V114/N57/assass.57w.html and http://articles.latimes.com/1995-06-30/news/mn-18914_1_white-house
9/12/94		1		Y	US	Frank Eugene Corder	Killed when he flew a stolen Cessna light plane into the White House south lawn.	Stolen aircraft	http://www.check-six.com/Crash_Sites/WhiteHouse_Corder.htm and http://www.nytimes.com/1994/09/13/us/crash-white-house-overview-unimpeded-intruder-crashes-plane-into-white-house.html
2/1/82		3	2		US	Frank G. Spisak, Jr.	Self proclaimed Nazi and racist. Convicted of thee murders. Supreme Court upheld death sentence. Executed in 2011.	Fire arms	http://murderpedia.org/male.S/s1/spisak-frank-v-smith.htm and https://executions2010.wordpress.com/tag/frank-g-spisak-jr/
Spring, 2001				Y	US	Hassan Abujihaad	U.S. Navy sailor convicted of supporting terrorism and disclosing classified information. Sentenced to 10 years in prison.	Information	http://www.fbi.gov/news/stories/2008/march/secrets031008 and http://www.justice.gov/usao-ct/us-v-hassan-abu-jihaad
6/29/05					Swiss	Herve Falcini	According to BBC, a whistle-blower. Swiss Attorney General charged him with breaking secrecy laws and industrial espionage. Suspected of taking financial records from HBSC.	IT	http://www.bbc.com/news/world-europe-31296007 and http://www.spiegel.de/international/business/interview-hsbc-swiss-bank-whistleblower-herve-falciani-on-tax-evasion-a-911279.html

Date	WMD	Killed	Wounded	Failed	Country	Perpetrator or Suspect	Circumstances	Weapons	Reference
7/4/02		3	4		US	Hesham Mohamed Hadayet	FBI named Hadayet as the El Al ticket stand shooter who killed 2 and wounded 4 people. Killed on site by security guard.	Fire arms, knife	http://edition.cnn.com/2002/US/07/04/la.airport.shooting/ and http://abcnews.go.com/US/story?id=91485
9/24/09				Y	US	Hosam Maher Husein Smadi	Arrested by FBI for attempting to use a **weapon of mass destruction**. Pleaded guilty to attempting to bomb a Dallas skyscraper. Pled guilty and sentenced to 24 years in prison	Bomb	http://www.fbi.gov/dallas/press-releases/2009/dl092409.htm and http://www.justice.gov/usao/txn/PressRel10/smadi_ple_pr.html
2/4/14		3	16		US	Ivan Lopez	Alleged Fort Hood Shooter; apparently disgruntled about being denied a longer leave period/ Lopez killed himself at the scene.	Fire arms	http://www.huffingtonpost.com/2014/04/07/ivan-lopez-requested-leav_n_5107121.html and http://www.washingtonpost.com/news/checkpoint/wp/2015/01/23/army-details-the-downward-spiral-of-the-fort-hood-shooter-ivan-lopez/
2/5/92		4	2		N. Ireland	James Allen Moore	According to the press, Moore was the prime suspect who was distraught over colleague's death IRA killed 3 people; found later, an apparent suicide	Fire arms	http://www.nytimes.com/1992/02/05/world/3-shot-dead-in-belfast-office-of-pro-ira-group.html and http://www.anphoblacht.com/contents/8416
7/20/12		12	70		US	James Eagan Holmes	Accused in Century movie theater shooting in Aurora, Colorado; a plea of insanity was accepted by the judge.	Fire arms	http://murderpedia.org/male.H/h/holmes-james.htm and http://news.yahoo.com/why-its-taken-so-long-to-bring-colorado-theater-shooter-james-holmes-to-trial-183657552.html
6/10/09		1			US	James von Brunn	United States Holocaust Memorial Museum shooting; von Brunn was accused of being a white supremacist and Holocaust denier. Charged with murder, he died awaiting trial.	Fire arms	http://www.washingtonpost.com/wp-dyn/content/article/2010/01/06/AR2010010604095.html and http://murderpedia.org/male.V/v/von-brunn-james.htm
8/1/12		6	13		US	Jared Loughner	Pleaded guilty to shooting Congresswoman Gabrielle Giffords and killing and wounding others at a political rally. Sentenced to life in prison.	Fire arms	http://www.cbsnews.com/news/jared-loughner-who-shot-gabrielle-giffords-in-tucson-ranted-online/ and http://articles.latimes.com/2012/nov/08/nation/la-na-nn-jared-loughner-life-in-prison-20121108

Date	WMD	Killed	Wounded	Failed	Country	Perpetrator or Suspect	Circumstances	Weapons	Reference
4/2/09		14	4		US	Jiverly Antares Wong	American Civic Association immigration center in Binghamton, New York. Possible motives: despair over poor English, loosing job, depression. Committed suicide after shootings.	Fire arms	http://en.wikipedia.org/wiki/Binghamton_shootings and http://murderpedia.org/male.W/w/wong-jiverly.htm
11/8/91 to Jan, 1992		1	11		Sweden	John Ausonius	Said to have had hatred for immigrants; all victims were immigrants. Convicted of murder and robbery. Later is said to have confessed to the serial shootings	Fire arms	http://murderpedia.org/male.A/a/ausonius-john.htm and http://www.findadeath.com/forum/showthread.php?18471-Swedish-killer-John-Ausonius-quot-The-Laser-Man-quot
10/14/12				Y	US	John F. Schrank	Attempted to murder President Theodore Roosevelt; a folded copy of his speech and a metal eye glass case in his pocket absorbed much of the force of the bullet's impact.	Fire arms	http://en.wikipedia.org/wiki/John_Flammang_Schrank and http://murderpedia.org/male.S/s/schrank-john.htm
3/30/81		1	2	Y	US	John Hinckley, Jr.	Assassination attempt on President Regan. White House press secretary James Brady was one of the wounded in 1981and his death in 2014 was ruled a homicide.	Fire arms	http://www.usatoday.com/story/news/nation/2014/08/04/james-brady-john-hinckley/13598699/ and http://www.washingtonpost.com/local/crime/prosecutors-will-not-charge-hinckley-with-murder-in-death-of-james-brady/2015/01/02/67de0024-929a-11e4-a900-9960214d4cd7_story.html
11/14/13	Y			Y	US	Jordan Gonzalez	According to FBI documents, during his guilty plea, Gonzalez admitted to "acquiring...materials...(for) developing ricin and abrin as weapons...in anticipation of using them in confrontations with other people in the future." The FBI said he also purchased material for manufacturing explosives. Sentenced to 78 months in prison.	Toxins, Explosives	http://www.fbi.gov/newyork/press-releases/2014/new-jersey-pharmacist-admits-attempting-to-weaponize-deadly-toxins-and-possessing-narcotics-manufacturing-equipment and http://www.justice.gov/opa/pr/pharmacist-sentenced-78-months-prison-attempting-weaponize-deadly-toxins-and-possessing

Date	W M D	Killed	Wou nded	Fail ed	Country	Perpetrator or Suspect	Circumstances	Weapons	Reference
11/20/1 1				Y	US	Jose Pimentel	New York Police Commissioner Raymond W. Kelly, and Mayor Michael R. Bloomberg said that Pimentel was planning to use pipe bombs to attack targets throughout New York City. Sting operation. Pled guilty and sentenced to 16 years in prison.	Pipe bombs	http://www.nytimes.com/2014/03/26/nyr egion/judge-imposes-16-year-term-for-manhattan-man-in-pipe-bomb-case.html?ref=topics and http://www.nydailynews.com/news/crim e/lone-wolf-terrorist-jose-pimentel-16-years-prison-article-1.1734472
4/18/10				Y	US	Joseph Brice	Brice pled guilty to manufacturing an unregistered explosive device, and attempting to provide material support to terrorists relating to explosives. Sting operation. Sentenced to 12 years	Bomb (threat)	http://news.intelwire.com/2011/06/jihadi st-who-wasnt.html and http://www.investigativeproject.org/cas e/613
1998-2001					US	Joseph Konopka	Called "Dr. Chaos/", Pleaded guilty to arson and other crimes .Received 13 year sentence which was overturned, But is serving out the sentence concurrently for hiding bottles of cyanide in Chicago's subway system.	IT	http://www.nndb.com/people/787/0000 97496/ and http://www.nbcnews.com/id/8055451/n s/us_news-crime_and_courts/t/dr-chaos-arson-conviction-overturned/#.VRtJjjvF_Gs
5/12/82		1		Y	Italy	Juan Maria Fernandez	Priest was found guilty of attempting to murder the Pope at Fatima; he is said to have disagreed with Pope's changing policies. Sentence 6 years, served three	Bayonet	http://www.nytimes.com/1983/05/03/wo rld/around-the-world-spanish-priest-sentenced-in-82-attack-on-pope.html and https://xenophilius.wordpress.com/200 8/10/16/juan-fernandez-krohn-1982-secret-pope-stabber/
3/9/11	Y			Y	US	Kevin William Harpham	Charged with planting a backpack containing a radio-controlled bomb along the planned route of a Martin Luther King Jr. Day march. Pleaded guilty to attempting to use a **weapon of mass destruction** and attempting injury because of race, color or national origin. 32 year sentence.	Bomb with fishing weights and rat poison	http://www.csmonitor.com/USA/Justice/ 2011/1220/Failed-Martin-Luther-King-Day-parade-bomber-gets-32-year-sentence and http://www.justice.gov/opa/pr/washingt on-man-sentenced-32-years-attempted-bombing-martin-luther-king-unity-march

Date	W M D	Killed	Wounded	Failed	Country	Perpetrator or Suspect	Circumstances	Weapons	Reference
2/23/11	Y			Y	US	Khalid Ali-M Aldawsari	Authorities said he acquired materials and researched targets to bomb including the Dallas residence of former President George W. Bush. He was convicted on charges of attempting to use of a **weapon of mass destruction** and sentenced to life imprisonment.	Car bombs	http://www.huffingtonpost.com/2012/06/27/khalid-alim-aldawsari-texas-bomb-plot_n_1630276.html and http://www.kcbd.com/story/20083237/lubbock-terror-plot-suspect-sentenced-to-life-in-prison
4/12/96		1	8		US	Larry Shoemake	Allegedly a white supremacist: who is said to have hated Jews, blacks and the government. Shooting spree in Jackson MI ended in his suicide	Fire arms	http://www.historycommons.org/entity.jsp?entity=larry_wayne_shoemake_1 and http://www.adl.org/assets/pdf/combating-hate/Explosion-of-Hate.pdf
11/22/63		1	1		US	Lee Harvey Oswald	Assassin of President John Kennedy	Fire arms	http://en.wikipedia.org/wiki/Assassination_of_John_F._Kennedy and http://www.history.com/this-day-in-history/jack-ruby-kills-lee-harvey-oswald
9/10/10			1	Y	Denmark	Lors Doukaiev	Danish police believed he was constructing a letter bomb that exploded during loading of explosive. Found guilty of attempted terrorism and sentenced to 12 years in prison.	TATP, letter bomb	http://en.wikipedia.org/wiki/Hotel_J%C3%B8rgensen_explosion and http://www.cnn.com/2011/WORLD/europe/05/30/denmark.attempted.terror.verdict/
5/3/02			6		US	Luke Helder	The Midwest Pipe Bomber; he was planting bombs in a smiley face pattern. Six of his 18 mailed bombs exploded. Charged for using an explosive device. Held incompetent to stand trial in 2004. In March 2012, a US District judge is quoted as saying that he will order a new competency hearing. (Sioux City Journal, May 18, 2003)	Pipe bombs	http://web.archive.org/web/20070926231308/http://www.usps.com/postalinspectors/ar02/ar02_04.htm and http://www.huffingtonpost.com/2013/05/15/luke-helder-mailbox-bomb-_n_3281576.html

Date	WMD	Killed	Wounded	Failed	Country	Perpetrator or Suspect	Circumstances	Weapons	Reference
Dec, 2005				Y	US	Michael C. Reynolds	Convicted of attempting to provide support to Al-Qaeda, soliciting a crime of violence, unlawful distribution of explosives, possession of a hand grenade. In an FBI sting, he is said to have been plotting to blow up a pipeline and an oil refinery.	Hand grenade, Explosives	http://www.foxnews.com/story/2006/10/04/pennsylvania-man-indicted-for-offering-to-help-al-qaeda/?sPage=fnc.specialsections/waronterror and http://www.investigativeproject.org/documents/case_docs/933.pdf
9/23/09	Y			Y	US	Michael Finton	Pleaded guilty of attempting to bomb. Springfield, IL Federal Building and Courthouse. In a plea agreement, was sentenced to 28 years in prison for attempted use of a **weapon of mass destruction**. FBI sting.	One ton car bomb	http://www.justice.gov/opa/pr/illinois-man-admits-plotting-bomb-federal-courthouse-and-sentenced-28-years-prison and http://www.huffingtonpost.com/2011/05/10/michael-finton-pleads-gui_n_859922.html
1/25/93		2	3		US	Mir Aimal Kansi (Kasi)	Admitted shooting CIA employees at Langley, VA as they sat in their cars. Reported motive: anger with US policy in Mid East. Fled country. Rendered. Tried and found guilty of murder. Executed 2002.	Fire arms	http://en.wikipedia.org/wiki/1993_shootings_at_CIA_Headquarters and http://www.sullivan-county.com/id3/kasi.htm
11/26/10	Y			Y	US	Mohamed Osman Mohamud	Attempted to detonate a car bomb at a Christmas tree lighting ceremony in Portland, Oregon. Charged with attempting to use a **weapon of mass destruction**. FBI sting. Sentenced to thirty years.	Car bomb	http://en.wikipedia.org/wiki/2010_Portland_car_bomb_plot and http://www.csmonitor.com/USA/USA-Update/2014/1001/Portland-Christmas-tree-bomber-gets-30-years-as-questions-about-arrest-linger
12/30/08				Y	US	Mohammed T. Alkaramla	Convicted of mailing bomb threat to a Jewish school; sentenced to two years in prison	Bomb (threat)	http://www.fbi.gov/chicago/press-releases/2009/cg032009a.htm and http://articles.chicagotribune.com/2010-11-24/news/ct-met-bomb-threat-sentencing-20101124_1_ida-crown-jewish-academy-jewish-high-school-threat

Date	WMD	Killed	Wounded	Failed	Country	Perpetrator or Suspect	Circumstances	Weapons	Reference
10/12/09		2			Italy	Mohammed Game	Attempted what is thought to have been a suicide attack on a Milan, Italy Italian military barracks. Severely injured when his IED went off but failed to cause wider destruction.	IED	https://www.ctc.usma.edu/posts/the-october-2009-terrorist-attack-in-italy-and-its-wider-implications and http://www.itnsource.com/shotlist//RTV/2009/10/14/RTV1955609/?v=1
1/1/10		1		Y	Denmark	Mohammed Geele	Sentenced to ten years in prison for attempted terrorism and manslaughter. He attempted to assassinate cartoonist Kurt Westergaard in Denmark.	Axe, knife	http://www.thenational.ae/news/world/europe/prophet-cartoonist-axe-attacker-guilty-of-terrorism and http://www.reuters.com/article/2011/02/04/idINIndia-54664020110204
3/11/12		8	5		France	Mohammed Merah	Killed Jewish schoolchildren, a Rabbi, and French paratroopers. Was killed in a standoff capture by French police	Fire arms	http://en.wikipedia.org/wiki/Toulouse_and_Montauban_shootings and http://www.huffingtonpost.com/2012/03/22/toulouse-shooting-mohamed-merah-shot-dead_n_1372387.html? d
3/14/06			9	Y	US	Mohammed Reza Taheri-azar	Sentenced to up to 30 years for driving an SUV into a crowd of students at the University of North Carolina. The District Attorney said his intent was to kill.	SUV	http://web.archive.org/web/20080502132128/http://www.newsobserver.com/102/story/415421.html and http://www.wral.com/news/local/story/3432689/
7/28/11	Y			Y	US	Naser Jason Abdo	Former US Army private convicted of attempted use of a **weapon of mass destruction**, attempted murder, and other charges. Sentenced to two consecutive life prison terms plus 60 years. Angry at Army; opposed war in Afghanistan	Fire arms	http://en.wikipedia.org/wiki/Naser_Jason_Abdo and http://www.fbi.gov/sanantonio/press-releases/2012/naser-jason-abdo-sentenced-to-life-in-federal-prison-in-connection-with-killeen-bomb-plot
7/28/06		1	5		US	Naveed Afzal Haq	Arrested for Seattle Jewish Federation shooting; cited anger at Israel and Judaism. Convicted of murder, hate crimes, and other crimes. Sentenced to life in prison for the murder followed by additional time for the other convictions.	Fire arms	http://www.nytimes.com/2006/07/30/us/30seattle.html and http://articles.latimes.com/2010/jan/15/nation/la-na-seattle-jewish-center15-2010jan15

Date	W M D	Killed	Wou nded	Fail ed	Country	Perpetrator or Suspect	Circumstances	Weapons	Reference
5/8/07				Y	UK	Nicholas Roddis	Charged with placing hoax bomb on bus. Received a 7 year sentence, After release, was accused of researching on how to make a real bomb, but cleared of plotting an attack.	Hoax bomb	http://www.dailymail.co.uk/news/article-2281909/Convicted-terrorist-claimed-using-bomb-making-chemicals-treat-warts-bad-teeth-cleared-plotting-attack.html and http://www.mirror.co.uk/news/uk-news/nicholas-roddis-convicted-terrorist-cleared-1721415
22 May, 2008		1		Y	UK	Nicky Reilly	Attempted to assemble three bombs in a restaurant; one blew up in his hands. Charged under the UK Terrorism act and Explosive Substances Act. Sentenced to life imprisonment (minimum 18 years).	Bomb	http://www.theguardian.com/uk/2008/oct/15/uksecurity and http://news.bbc.co.uk/2/hi/uk_news/7859887.stm
Nov 5 2009		13	30		US	Nidal Hasan	Army psychiatrist convicted of premeditated murder and attempted murder in the Fort Hood mass shooting. Sentenced to death;	Fire arms	http://en.wikipedia.org/wiki/Nidal_Malik_Hasan and http://www.washingtonpost.com/world/national-security/nidal-hasan-sentenced-to-death-for-fort-hood-shooting-rampage/2013/08/28/aad28de2-0ffa-11e3-bdf6-e4fc677d94a1_story.html
5/19/10				Y	US	Paul G. Rockwood, Jr.	Pleaded guilty to charges that he made false statements to the FBI about his list of 15 individuals to be targeted for violent attacks. Sentenced to 8 tears in prison.	Firearm s, Mail bombs	http://www.justice.gov/archive/usao/ak/news/2010/August/Rockwood_Paul_08-24-10.html and http://www.fbi.gov/news/stories/2012/november/in-alaska-a-domestic-terrorist-with-a-deadly-plan
7/13/03 and 10/10/09		2	5		Sweden	Peter Mangs	15 suspected sniper shootings. Xenophobic. Accused of shooting at immigrants among others. Sentenced to life in prison.	Fire arms	http://murderpedia.org/male.M/m/mangs-peter.htm and http://www.thelocal.se/20121123/44606
2/15/10				Y	UK	Rajib Karim	Convicted of preparing acts of terrorism. Accused of planning to obtain information to down a trans-Atlantic jet, or induce others at BA to load a package on a plane bound for the US. Sentenced to 30 years in prison.	IT	http://www.bbc.com/news/uk-12788224 and http://www.theguardian.com/uk/2011/feb/28/british-airways-bomb-guilty-karim

Date	W M D	Killed	Wou nded	Fail ed	Country	Perpetrator or Suspect	Circumstances	Weapons	Reference
4/11/96				Y	US	Ray Hamblin	Hamblin, described as a survivalist, was arrested when ATF agents found a large cache of explosives and weapons on his property. Sentenced to almost 4 years in prison, and released..	Bombs, grenade s	https://news.google.com/newspapers?nid=2199&dat=19960406&id=sqcyAAAAIBAJ&sjid=4uYFAAAAIBAJ&pg=5173,1524170&hl=en and http://www.splcenter.org/get-informed/publications/terror-from-the-right
9/28/11				Y	US	Rezwan Ferdaus	Pleaded guilty to attempting to damage and destroy a federal building; accused of planning an attack on the Pentagon and Capitol Building using a small drone aircraft laden with explosives. Sentenced to 17 years in prison.	Drone, Cellpho ne detonat or	http://www.nytimes.com/2012/11/02/us/rezwan-ferdaus-of-massachusetts-gets-17-years-in-terrorist-plot.html and http://www.boston.com/metrodesk/2012/11/01/rezwan-ferdaus-ashland-sentenced-years-terror-plot/KKvy6D6n2PfXfbEfA4iMwJ/story.html
12/22/01	Y			Y	US	Richard Reid	Shoe Bomber, attempted to destroy an international airliner. Convicted of attempted use of a **weapon of mass destruction** and other charges. Was sentenced to three consecutive life terms plus 110 years in prison.	PETN and TATP, Plastic explosiv es	http://en.wikipedia.org/wiki/Richard_Reid and http://www.dailymail.co.uk/news/article-2047093/Shoe-bomber-Richard-Reid-pictured-inside-US-Supermax-jail.html
7/22/96		2	111		US	Robert Rudolph	Atlanta Olympic Park serial bomber. white supremacist, anti gay and anti-abortion. Convicted of the Olympic Park bombing and sentenced to four consecutive life terms plus 120 years.	Bombs	http://www.cnn.com/2013/09/18/us/olympic-park-bombing-fast-facts/index.html and http://murderpedia.org/male.R/r/rudolph-eric.htm
12/16/89		2	1		US	Roy Moody	Convicted of all counts associated with mailed bombs including murder of a judge and a civil rights attorney killed by the bombs. In prison on death row	Mail bombs	http://www.unabombers.com/VanPac/v91-06-29-Moodytrial-3.htm and http://blog.al.com/spotnews/2013/02/convicted_pipe_bomber_walter_l.html

Date	WMD	Killed	Wou nded	Fail ed	Country	Perpetrator or Suspect	Circumstances	Weapons	Reference
1/27/12	Y			Y	US	Sami Osmakac	Convicted of plotting terrorist attacks; he had described his plans to bomb and kill people in a video Apprehended after he sought to buy a Al-Qaeda flag. Protested treatment of Muslims. Sting operation. Convicted of planning to use a **weapon of mass destruction.** Sentenced to 40 years in prison.	Suicide vest	http://www.tampabay.com/news/courts/criminal/sami-osmakac-gets-40-years-in-prison-for-plotting-terrorist-attacks-in/2205214 and http://www.investigativeproject.org/case/608
9/19/10	Y			Y	US	Sami Samir Hassoun	Pleaded guilty to charges that he attempted to use of a **weapon of mass destruction** and explosive device. Believed to have placed a backpack containing what he though was a bomb in trash can outside of Wrigley Field; it was inert. Sting operation. Sentenced to 23 years in prison.	Backpac k bomb	http://www.dailymail.co.uk/news/article-2076937/Sami-Samir-Hassoun-Terrorist-planted-fake-bomb-bin-outside-Wrigley-Field.html and http://www.huffingtonpost.com/2013/05/30/sami-samir-hassoun-senten_n_3359234.html
2/22/74	Y		1	Y	US	Samuel Joseph Byck	Attempted to hijack a DC-9 and thought to have planned to fly it into the White House to kill President Nixon; shot pilots but was killed himself when the plane on the tarmac was stormed.	Hi jacked aircraft	http://www.todayifoundout.com/index.php/2012/02/this-day-in-history-samuel-byck-hijacks-an-airliner-with-the-intent-of-flying-it-into-the-white-house-to-kill-president-nixon/ and http://murderpedia.org/male.B/b/byck-samuel.htm
5/31/09		1			US	Scott Roeder	Anti-abortion activist convicted of killing an abortion doctor, Dr. George Tiller. Fifty year sentence without parole.	Fire arms	http://murderpedia.org/male.R/r/roeder-scott.htm and http://www.kansas.com/news/article1033388.html
4/16/07		32	17		US	Seung-Hui Cho	Virginia Polytechnic Institute shooter. Committed suicide after the shootings.	Fire arms	http://abcnews.go.com/US/story?id=3048108 and http://www.washingtonpost.com/wp-dyn/content/article/2007/04/18/AR2007041800162.html
6/20/13				Y	Indonesi a	Sefa Riano	Allegedly planning to bomb the Myanmar Embassy; published plans on Facebook. Five pipe bombs found in backpack.	Pipe bombs	http://www.huffingtonpost.com/2013/06/20/sefa-riano-caught-through-facebook-indonesia_n_3471594.html http://www.arabnews.com/news/450314 and

Date	W M D	Killed	Wou nded	Fail ed	Country	Perpetrator or Suspect	Circumstances	Weapons	Reference
2/14/08		5	21		US	Steven Kazmierczak	Shooter in the Northern Illinois University shooting. Wrote a paper on Hamas and its social service projects. Committed suicide after mass killings.	Fire arms	http://en.wikipedia.org/wiki/Northern_Illinois_University_shooting and http://murderpedia.org/male.K/k/kazmierczak.htm
12/11/10		1	2	Y	Sweden	Taimour Abdulwahab al-Abdaly	Prosecutor said police were 99% sure that al-Abdaly was the suicide car bomber. Bomb apparently exploded prematurely. In Stockholm.	Car bomb	http://www.bbc.com/news/world-europe-11983667 and http://www.telegraph.co.uk/news/uknews/terrorism-in-the-uk/8198043/Sweden-suicide-bomber-Taimur-Abdulwahab-al-Abdaly-was-living-in-Britain.html
1978-1995		3	23		US	Theodore Kaczynski	The "Unabomber", anti-technology Luddite. Pleaded guilty and sentenced to life in prison without parole.	Mailed bombs	http://www.crimemuseum.org/crime-library/ted-kaczynski-the-unabomber and http://www.encyclopedia.com/topic/Unabomber.aspx
4/19/95	Y	168	600		US	Timothy McVeigh	Oklahoma City bombing. Motive was apparently retaliation for the Waco Siege, Ruby Ridge, other government raids and general U.S. foreign policy. Found guilty and executed in June 2011.	Truck bomb	http://en.wikipedia.org/wiki/Timothy_McVeigh and http://law2.umkc.edu/faculty/projects/ftrials/mcveigh/mcveighsentencing.html
12/25/09	Y			Y	US	Umar Farouk Abdulmutalla	The Underwear Bomber, attempted to bring down an airliner. Pled guilty to commit an act of terrorism and attempting to use a **weapon of mass destruction**. Sentenced to life in prison.	PETN and TATP, Plastic explosives	http://www.justice.gov/opa/pr/umar-farouk-abdulmutallab-sentenced-life-prison-attempted-bombing-flight-253-christmas-day and http://www.theguardian.com/world/2012/feb/16/underwear-bomber-sentenced-life-prison
11/4/95		1	1		Israel	Yigal Amir	Assassinated Israele premier Yitzhak Rabin. Serving a life sentenze plus extra time for the killing.	Fire arms	http://en.wikipedia.org/wiki/Yigal_Amir and https://www.jewishvirtuallibrary.org/jsource/biography/Yigal_Amir.html
6/17/11				Y	US	Yonathan Melaku	Accused of firing at military buildings, in three separate attacks including the Pentagon. Picked up at Arlington National Cemetery and charged. Plea deal resulted in a sentence of 25 years in prison.	Fire arms	http://www.washingtonpost.com/local/crime/pentagon-shooter-sentenced-to-25-years-in-prison/2013/01/11/a9fbd47a-5c2a-11e2-9fa9-5fbdc9530eb9_story.html and http://www.fbi.gov/washingtondc/press-releases/2013/alexandria-man-sentenced-to-25-years-for-shooting-military-buildings-in-northern-virginia

Date	W M D	Killed	Wou nded	Fail ed	Country	Perpetrator or Suspect	Circumstances	Weapons	Reference
7/10/10				Y	US	Zachary Chesser	Arrested before he flew to Somalia. He had threatened South Park animators for using what he considered sacred images. Pleaded guilty to communicating threats, to providing material support to a designated terrorist organization and other charges. Sentenced to 25 years in prison.	Threats and intimidat ion	http://en.wikipedia.org/wiki/Zachary_Ad am_Chesser and http://www.justice.gov/opa/pr/virginia-man-sentenced-25-years-prison-providing-material-support-and-encouraging-violent
8/24/14		1	3		US	Zale Thompson	Suspected in New York City hatchet attacks that injured 2 policemen. Killed on the scene by police.	Hatchet	http://www.inquisitr.com/1559728/nyc-police-hatchet-attack-suspect-zale-thompson-may-have-ties-to-isis/ and http://nypost.com/2014/10/23/man-shot-dead-after-striking-cop-in-the-head-with-ax/
1/21/15			12		Israel	Hamza Muhammad Hassan Matruch,	Reported to have stabbed people aboard a public bus, shot in leg by police	Knife	http://www.jpost.com/Israel-News/Three-stabbed-on-Tel-Aviv-bus-388417 and http://www.unitedjerusalem.org/index2. asp?id=1864032
2/15/15		2	5		Denmar k	Suspect: Omar Abdel Mamid El-Hussein	Reported to have shot people at a bar mitzvah ceremony in a synagogue	Fire arms	http://en.wikipedia.org/wiki/2015_Cope nhagen_shootings and http://www.cbsnews.com/news/copenh agen-gunman-omar-abdel-hamid-el-hussein-just-out-of-prison-ap-sources/

Appendix F: Authors Biographies

Theodore Jay Gordon is a futurist and management consultant, a specialist in forecasting methodology, planning, and policy analysis. He is co-founder and Board member of The Millennium Project, a global think tank. He formed the consulting firm The Futures Group in 1971 and led it for 20 years. Prior to forming The Futures Group, he was one of the founders of The Institute for the Future and before that consulted for the math and policy department at RAND Corporation. He was also Chief Engineer of the McDonnell Douglas Saturn S-IV space vehicle and was in charge of the launch of early ballistic missiles and space vehicles from Cape Canaveral. He is a frequent lecturer, the author of many technical papers and author or co-author of several books dealing with space, the future, life extension, and scientific and technological developments and issues. He is the author of the Macmillan encyclopedia article on the future of science and technology. He is currently on the editorial board of *Technological Forecasting and Social Change*.

Dr. Yair Sharan, a Col(ret) in the IDF, is currently the director of TAM-C/FIRST group active in the security and technology field. He has been a senior associate researcher in Begin-Saadat Center for strategic studies (BESA) in Bar Ilan University in Israel. He was the director of the Interdisciplinary Center for Technological Analysis and Forecasting (ICTAF) at Tel Aviv University, Israel, in the period 2000-2012 and was a senior consultant to the Israeli Ministry of Science in 1992–1993. Dr. Sharan served as Israeli Science Counselor in Germany from 1988 to1992 and coordinated the relations of the University of Tel-Aviv with the EU in the period 1993–2003. Dr. Sharan is co-director of the Israeli Node of the Millennium Project, a member of the World Future Society and has been a member of the EU Foresight Expert Group.

His main fields of interest are research and science policy, the impact of technology on national strategy, technology foresight, and security foresight, technology assessment, and more. He has special expertise in the security field and has completed studies in topics like "Non Conventional Terrorism", "Psychological Deterrence", "Emerging Technologies and Future Threats of Terrorism", "Issues in water security", and more. He is co-editor of three books including "Terrorism on the Internet", "Medical Response to Emergency Situations" and "Water Security in Emergency Situations." He has been co-director of several NATO SPS workshops, the most recent one on Lone Actor terrorism, in November 2014. Dr. Sharan coordinated several EU projects, including FESTOS in the Security program and PRACTIS in the Science in Society program. Together with colleague researchers he participated in many EU projects including SIAM, e-Leaving, OPET, Nano2Life, Platform Foresight, and others. Current studies and publications include "Lonely wolf--an emerging terror threat," "Synthetic Biology--Risks and Chances," and "Emerging technologies and their impact on security and privacy."

Elizabeth Florescu is Director of Research at The Millennium Project, working with the Project since 1997. She is co-author of the annual "State of the Future", the report "Environmental Crimes, Military Actions, and the International Criminal Court", the "Analysis of the UN Millennium Summit Speeches", several articles, and invited reviewer at specialized magazines. She has been one of the principal investigators working on the monthly environmental scanning reports for the US Army Environmental Policy Institute, assessing worldwide environment and new technologies-related issues that might trigger future changes to international regulations with potential implications for the military. Elizabeth has been also actively participating in the design of a climate change situation room, design and implementation of early warning systems for policymakers, and building a collective intelligence system, and is involved in committees and forums addressing issues related to S&T, environment, security, international regulations, ontology, and knowledge management. Elizabeth has a vast international experience, having lived and worked in several countries, including the U.S., Canada, Romania, and Hungary.

List of Figures and Tables

In parenthesis are indicated the number of the page of the respective figure or table.

Acronyms and Abbreviations

AI	Artificial Intelligence
ATF	Bureau of Alcohol, Tobacco, Firearms and Explosives (U.S.A.)
BWC	Biological Weapons Convention
CBW	Chemical and Biological Weapons
CDC	Center for Disease Control
CIS	Collective Intelligence System
CRG	Cyber Response Group
CTITF	Counter-Terrorism Implementation Task Force Office
CWC	Chemical Weapons Convention
DARPA	Defense Advanced Research Projects Agency (U.S.A.)
DEC	Digital Equipment Corporation
DDoS	Distributed Denial of Services
DIY	do it yourself
DNA	deoxyribonucleic acid
EC	European Commission
EEG	Electroencephalography
ETC Group	Erosion, Technology and Concentration action group
EU	European Union
FAO	Food and Agriculture Organization of the UN
FBI	Federal Bureau of Investigation (U.S.A.)
FDA	Food and Drug Administration (U.S.A.)
FESTOS	Foresight of Evolving Security Threats Posed by Emerging Technologies
FMD	foot and mouth disease
fMRI	functional magnetic resonance imaging
GFIS	Global Futures Intelligence System
GM/GMO	Genetically modified organism
GTB	Global Terrorism Database
IAEA	International Atomic Energy Agency
IASB	International Association of Synthetic Biology
ICT	information and communication technology
IEA	International Energy Agency
IED	improvised explosive device
IoT	Internet-of-Things
iPS	induced Pluripotent Stem Cells
IQ	intelligence quotient
IS	Islamic State (or Islamic State of Iraq and the Levant)

ISIS	Islamic State of Iraq and Syria (or the Islamic State of Iraq and ash-Sham)
ISPs	Internet service providers
LFW	Labeled Faces in the Wild
IVF	in vitro fertilization
LW	lone wolf
MAOA	monoamine oxidase-A gene
MENA	Middle East and North Africa
MIT	Massachusetts Institute of Technology
NASA	National Aeronautics and Space Administration (U.S.A.)
NATO	North Atlantic Treaty Organization
NBC	Nuclear, Biological, and Chemical
NEMS	nano-electromechanical systems
NGO	Nongovernmental organization
NIH	National Institutes of Health
NT	nanotechnology
NSA	National Security Agency
NYU	New York University
OECD	Organisation for Economic Co-operation and Development
OPCW	Organisation for the Prohibition of Chemical Weapons
PCR	Polymerase Chain Reaction
PET	positron emission tomography
PKK	Kurdistan Workers' Party (Partiya Karkerên Kurdistanê)
PPE	personal protective equipment
PTSD	post-traumatic stress disorder
R&D	research and development
RNA	ribonucleic acid
RTD	Real-Time Delphi
S&T	Science and technology
SIMAD	Single individual massively destructive
SIPRI	Stockholm International Peace Research Institute
SUV	sport utility vehicle
FIRST	Foresight Insight Research Science and Technology
TAO	Tailored Access Operation
TIC	Toxic Industrial Chemicals
TOC	Transnational organized crime
U.K.	United Kingdom
UKUSA	United Kingdom – United States of America Agreement (multilateral agreement for cooperation in signals intelligence between the UK, U.S.A., Canada, Australia, and New Zealand)
UNDP	United Nations Development Programme

UNFPA	United Nations Population Fund
UNHCR	United Nations High Commission on Refugees
UNODC	UN Office on Drugs and Crime
U.S.	United States
WHO	World Health Organization
WMD	weapon of mass destruction

INDEX

best anti-LW practices · 107
Bhopal India · 42
Big Brother · 68, 114
big data · 64, 98
big databases · 45
biocompatible implants · 26
Bio-engineers · 20
Bio-hacking · 22
biological · 3, 4, 5, 12, 19, 20, 22, 23, 29, 31, 32, 35, 36, 37, 42, 43, 44, 46, 52, 68, 70, 72, 77, 84, 104, 106, 132, 142
Biological weapons · 43
Biological Weapons Convention · 43, 100
biomarkers · 23
biomimicking · 21
Biotechnolog · 24
Biotechnology · 19
bio-threats · 25
bioweapon stockpiles · 43
Bitcoins · 72
black hats · 48
blackmail · 59, 61, 112
Bob Thomas · 53
Bolt, Beranek and Newman · 53
bomb making materials awareness program (BMAP), · 70
Boston Marathon Bombers · 3, 69
Boston Marathon killings · 164
Botulism · 43
brain imaging · 5, 12, 81, 106, 126, 137
brain scan · 82
brain scanning · 12, 107
Brazil · 102
BRCA1 · 23
breakthroughs · 18
Bruce Ivins · 20, 164
Brunei · 58
Buford O. Furrow, Jr. · 164
bulldozer · 3, 4

C

camouflage · 27, 29, 37
Canada · 10, 58, 98, 101, 102, 134, 159
Capitol Building · 63, 173
CBW · 42
censorship · 87, 97
Charlie Coulibaly · 97
Charlie Hebdo · 95, 97
chat rooms · 63, 99
chemical weapons · 43, 44
Chemical Weapons Convention · 4

Chemical Weapons Convention (CWC) · 42
child pornography · 71, 96, 97
chimeras · 23
China · 10, 24, 54, 61, 102, 103, 133
Chinese military hackers · 61
Chinese or other hackers · 99
chlorine · 42
Chris White · 60
Christopher Cornell · 73, 163
cilia · 21
civil liberties · 5, 57, 102, 105, 106, 107, 145
climate of trust · 92
Clint Eastwood · 86
cloaking · 37
CNN · 5, 87
code of conduct · 45
Columbine · 85, 86
Comprehensive Convention on International Terrorism · 101
compromises to freedoms · 114
computer · 20, 22, 23, 27, 34, 39, 49, 51, 53, 55, 59, 60, 61, 67, 68, 71, 72, 76, 81, 86, 107, 110, 113, 157
confidence · 6, 8, 10, 11, 12, 13, 14, 16, 17, 106, 108, 127, 128, 130, 134, 139
confidentiality privileges denied · 104
contrary marketing campaign · 52
control system of aircraft in flight · 112
Conventional defense and legal structures · 94
copycat · 85, 86, 87, 105
copycats · 48
corruption · 101, 150
Countermeasures · 103
counterterrorism · 93, 94, 159
Craig Venter · 20
crimes against humanity · 101
cross-border co-operation between law enforcement and related organizations · 100
crowdsourced · 114
cryptography · 55
Cryptolocker malware · 55
cultural · 94, 117
cyanide · 51, 87, 168
cyber attacks · 54, 90, 111, 112
cyber barriers around crucial systems · 113
cyber command · 60, 76
Cyber Hat · 55
Cyber operations · 111
Cyber Response Group (CRG) · 60
cyber systems · 107, 113
Cyber Threat Intelligence Integration Center · 60
cyber-Pearl-Harbor · 54
cybersecurity · 98
Cyberspace · 48, 54, 55, 90, 92, 102

Cyberweapons · 53
Cyborg insects · 33
Cylance · 112
Czech Republic · 58

D

Dairy Queen · 99
Dale Selby Pierre · 86
Dallas · 166, 168
Dan Kaufman · 59
darknet · 60, 72, 73, 96
DARPA · 33, 36, 37, 38, 59, 60, 156
data breaches · 99
databases · 24, 64, 65, 66, 67, 68, 99, 115
David Copeland · 165
David Mobilio · 88
DeepFace · 67
defense R&D · 27
demographics · 2
Denmark · 78, 169, 170, 176
Department of Homeland Security · 52, 70, 77, 79
Derrick Shareef · 165
designer babies · 24
designer embryos · 22
detection · 1, 2, 5, 12, 21, 31, 39, 45, 46, 57, 66, 67, 70,
    76, 80, 106, 107, 110, 111, 114, 115, 117, 126, 137,
    152, 158
Devin Moore · 86, 165
Dieudonné M'bala M'bala · 97
Director of National Intelligence · 60, 99
Director of the Counter-Terrorism Implementation Task
    Force Office (CTITF) · 101
disagreement · 3, 8, 81, 106, 108, 130
Distributed Denial of Services · 54
distrust among allies · 98
DIY biology · 20
DIY synthetic biology · 21
DNA · 21, 22
DNA sequencing · 21
do it yourself manuals · 95
Doctor Chaos · 51
Doom · 86
drone(s) · 3, 32, 102, 116
DROPOUTJEEP · 58
drug dealing · 71, 75
drug delivery systems · 26
drug development · 19
drugs · 23
Dylan Klebold · 86
Dzhokhar and Tamerian Tsamnaev · 3, 165

E

early warning · 1, 5, 17, 33, 48, 138
eavesdrop · 55
eavesdropping · 32, 57, 58, 111, 138
Ebola · 79, 94
Eden Natan-Zada · 165
Edward Snowden · 4, 49, 98, 165
Egyp · 43
electroencephalography · 81, 114
Electroencephalography (EEG) · 81
Electronic Communications Privacy Act · 102
Electronic Crimes Task Force · 61
email · 57
Emerson Winfield Begolly · 165
Empire State Building · 162
Encryption · 55
enforcement · 48, 64, 71, 72, 75, 76, 77, 100, 101, 102,
    104, 112, 116, 147, 161
environment · 18, 19, 24, 26, 37, 45, 54, 94, 104, 106,
    110, 136, 151, 153
epidemic · 4, 9, 43, 130
epidemiology · 25
Eric Harris · 86
Eric Klein · 2
Eric Rudolph · 165
Espionage · 49, 54, 61, 165
Estonia · 58
ethical dilemmas · 5
Ethiopia · 58
eugenics · 24
Europe · 10, 30, 50, 91, 92, 96, 132, 133, 134, 147
European Commission · 18, 26, 94
Everett Stern · 50
extremist ideology · 96

F

Facebook · 55, 64, 65, 66, 67, 73, 74, 97, 99, 175
Faisal Shahzad · 4, 5, 77, 166
false negatives · 68, 80, 92, 104, 105
false positives · 44, 68, 80, 92, 105
Farooque Ahmed · 166
FASCIA · 57
FAST system · 79
FBI · 20
FBI informant · 73
FBI sting · 162, 164, 166, 170, 171
Federal Bureau of Investigation · 20, 70, 77, 4
fertilizer · 4, 69
FESTOS · 2, 18, 19, 25, 29, 34, 40, 41, 109

film violence · 85
Financial Crimes Enforcement Network · 76
Finland · 84
FinSpy, FinFisher · 58
FIRST · 2
First Amendment rights · 76
Five Eyes · 98
fMRI · 5, 12, 81, 82, 106, 107, 114, 151
food · 24
food supply · 24
Foreign Intelligence Surveillance Act · 100
forensic police work · 29
foresight studies · 46, 110
Fort Hood · 62, 74, 91, 159, 167, 173
Fourth Amendment · 68
Fragile States Index · 93
France · 4, 50, 78, 96, 97, 171
Francisco Martin Duran · 166
Frank Eugene Corder · 166
Frank G. Spisak, Jr. · 166
freedom of publication · 109
freedom of speech · 104, 144

G

Gabriel Weimann · 52, 91
gasoline · 4
Gaza · 163
Gene amplification · 21
Gene therapy · 24
Gene Transfer · 22, 26
genetic analysis · 106, 114
genetic components · 20
genetic engineering · 19, 20, 46
genetic forensics · 21
genetically-modified crop varieties · 24
genetics · 19, 23, 83, 84, 106
Geneva Convention should be updated · 100
genocide · 101
geo-fencing · 75
Geoffrey Miller · 24
George Tiller · 174
Germany · 58, 78, 163
global networks · 18
Global Terrorism Database (GTB) · 25
Global Terrorist Attacks Database · 91
glocalized · 116
glocolization · 93
Goodwill Industries · 99
Google · 75, 99, 129
Grand Theft Auto · 86, 166
Grand Theft Auto Vicecity · 86

Grey Goo Scenario · 27

H

hackers · 48
Hacktivists · 111
hactivists · 71
Hamza Muhammad Hassan Matruch · 176
hand grenade · 170
handwriting or linguistic analysis · 112
Hare Psychopathy checklist · 80
Hassan Abujihaad · 166
hazardous materials · 28
health records · 45, 64, 66
healthcare networks · 112
HeartBleed · 55
Helfgott · 85, 87
Hello Everyone, my name's Andy · 88, 163
hemorrhagic fever · 43
Heritage Foundation · 63
Herve Falciani · 49
Herve Falcini · 166
Hesham Mohamed Hadayet · 167
Hezbollah · 51, 101
high-resolution video camera payload · 103
Home Depot · 69, 99
honeypot · 73
horizon scanning · 110
Hosam Maher Husein Smadi · 166
How to Make a Bomb in Your Mom's Kitchen · 74
HSBC · 49, 50
human rights · 101, 102, 109, 115, 153
hybrid plants · 23

I

I am Charlie · 97
identify theft · 59
IED · 5, 38, 69, 70, 160, 164, 171
If you see something, say something · 77
illegal products · 71
immigrants · 168, 173
improvised bombs · 3
incendiary devices · 68
incentive for action · 114
India · 58
Indonesia · 58
infectious agents · 21
InfoChemistry · 38
infrastructure · 3, 4, 18, 34, 40, 53, 99, 106, 113, 132, 136

T

U

Theodore J Gordon tedjgordon@gmail.com
Elizabeth Florescu Elizabeth@millennium-project.org
Yair Sharon sharany@gmail.com

2015

www.ingramcontent.com/pod-product-compliance
Lightning Source LLC
Chambersburg PA
CBHW061736270326
41928CB00011B/2255